The Time Travel Handbook

A Manual of Practical Teleportation & Time Travel

edited by

David Hatcher Childress

The **New Science Series**:
- THE TIME TRAVEL HANDBOOK
- THE FREE ENERGY DEVICE HANDBOOK
- THE FANTASTIC INVENTIONS OF NIKOLA TESLA
- THE ANTI-GRAVITY HANDBOOK
- ANTI-GRAVITY & THE WORLD GRID
- ANTI-GRAVITY & THE UNIFIED FIELD
- ETHER TECHNOLOGY
- THE ENERGY GRID
- THE BRIDGE TO INFINITY
- THE HARMONIC CONQUEST OF SPACE
- VIMANA AIRCRAFT OF ANCIENT INDIA & ATLANTIS
- UFOS & ANTI-GRAVITY: Piece For a Jig-Saw
- THE COSMIC MATRIX: Piece For a Jig-Saw, Part II

The **Mystic Traveller Series:**
- IN SECRET TIBET by Theodore Illion (1937)
- DARKNESS OVER TIBET by Theodore Illion (1938)
- IN SECRET MONGOLIA by Henning Haslund (1934)
- MEN AND GODS IN MONGOLIA by Henning Haslund (1935)
- MYSTERY CITIES OF THE MAYAby Thomas Gann (1925)
- THE MYSTERY OF EASTER ISLAND by Katherine Routledge (1919)
- SECRET CITIES OF OLD SOUTH AMERICA by Harold Wilkins (1952)

The **Lost Cities Series**:
- LOST CITIES OF ATLANTIS, ANCIENT EUROPE & THE MEDITERRANEAN
- LOST CITIES OF NORTH & CENTRAL AMERICA
- LOST CITIES & ANCIENT MYSTERIES OF SOUTH AMERICA
- LOST CITIES OF ANCIENT LEMURIA & THE PACIFIC
- LOST CITIES & ANCIENT MYSTERIES OF AFRICA & ARABIA
- LOST CITIES OF CHINA, CENTRAL ASIA & INDIA

The **Atlantis Reprint Series:**
- THE HISTORY OF ATLANTIS by Lewis Spence (1926)
- ATLANTIS IN SPAIN by Elena Whishaw (1929)
- RIDDLE OF THE PACIFIC by John MacMillan Brown (1924)
- THE SHADOW OF ATLANTIS by Col. A. Braghine (1940)
- ATLANTIS MOTHER OF EMPIRES by R. Stacy-Judd (1939)

The Time Travel Handbook

edited by
David Hatcher Childress

The Time Travel Handbook

edited by David Hatcher Childress

Art Direction by Eliot R. Brown

First Printing

July 1999

ISBN 0-932813-68-2

Printed in the United States of America

Published by
Adventures Unlimited Press
One Adventure Place
Kempton, Illinois 60946 USA
auphq@frontiernet.net

10 9 8 7 6 5 4

The Time Travel Handbook

CONTENTS

Acknowledgements

Many thanks to the following authors who helped this author, including Morris K. Jessup, Carl Allen, Jack Parsons, Ivan T. Sanderson, William L. Moore, Charles Berlitz, J. Manson Valentine, Alfred Bielek, Preston Nichols, Peter Moon, Arthur T. Winfree, John Gribbon, Eliot R. Brown, Paul J. Nahin, William Corliss, Jerome Clark, Brad Steiger, Timothy Beckley, Michael Talbot, Riley Crabb of Borderland Sciences, and many others.
May your good works be recognized from time to time.

Dedication

To Jennifer, who helped make this book a reality and whom I want to spend forever in time with.

The Time Travel Handbook

edited by
David Hatcher Childress

Illustration by Mick Brownfield

1.

The Fundamentals of Time Travel

From Quantum Mechanics to Superstring Theories

Time travels in divers paces with divers persons.
I'll tell you who Time ambles withal,
who Time trots withal,
who Time gallops withal,
and who he stands still withal.
—William Shakespeare

The consciousness of man has an "extended" nature,
which enables him to surpass the ordinary bounds of space and time
—suggesting that there is another dimension beyond the material world.
—Edgar Mitchell, former astronaut

Historical Theories on Time and Gravity

Welcome to Time Travel 101. Throughout history there have been many attempts to describe or explain gravitation, as well as the phenomenon of "time," sometimes referred to ancient philosophical texts as "the arrow of time." In order to gain a full understanding of time travel, it is important to grasp historical theories on time, gravity, and what is now termed the "Unified Field."

The various physical forces surrounding us are still little-understood. Each year brings us closer to understanding the amazing universe around us and how it works, the interaction with space-time and the so-called quantum field that is everywhere. Finding hyperspace wormholes and other time tunneling devices are the natural product of understanding how the universe works—and a strange universe it is!

Although Western society has gone through a number of dark ages, as far back as 330 B.C., Aristotle claimed that the four elements—earth, water, air, and fire—have their natural places, toward which they tend to travel. He argued that objects containing greater amounts of earth than others would fall toward the Earth faster and that their speed would increase as they neared their natural place.

Galileo Galilei deduced (1604) that gravity imparts a definite acceleration, rather than a velocity, and that this acceleration is the same for all objects traveling in a vacuum. The universality of gravitational acceleration is known as the weak equivalence principle.

Sir Isaac Newton made the most significant contribution to gravitational theory when he perceived (1606) that the orbit of the Moon depended on the same type of force that causes an apple to fall to Earth. This proposition required that the magnitude of the force decrease in inverse proportion to the square of the distance from the Earth's center. Newton combined the inverse square law with his three laws of motion to formulate a theory of universal gravitation, which stated that there is a gravitational attraction between every pair of objects, inversely proportional to the square of the distance between them.

Rene Descartes (1596-1650) had earlier proposed a nonquantitative theory of gravitation based on the inward pressure of vortices on planets, but Newton did not offer a cause for the attraction. In fact, he avoided even calling it an attraction, speaking instead of "bodies gravitating towards one another." This description was sufficient to deduce Kepler's laws of planetary motion, the oceanic tides, and the precession of the equinoxes. In 1846 it was used to predict and discover a new planet, Neptune. Thus Newton's theory of gravitation stands as one of the greatest advancements of scientific knowledge.

Expressed mathematically, Newton's theory states that there is an attractive force F, given by F = Gm(1)m(2)/r(2), between two particles having masses m(1) and m(2) and separation distance r. G is the gravitational constant of proportionality, an unknown quantity that could not be determined by solar-system observations, which give only ratios of masses and thus the product of G and some mass. The value of G was first determined in 1798 by Henry Cavendish, who measured the force of gravitational attraction between two spheres of known mass. This experiment has come to be known as "weighing the Earth," because once G was determined, the Earth's mass, m(e1), could be determined from the astronomically known value for Gm(e). The experiment has been repeated many times with increasing precision. The currently accepted value of G is

6.67259 X 10(-8) cm(3)/g sec(2).

Modern Theories

In 1905, Albert Einstein developed his theory of special relativity, which modified Newton's theory of gravitation. Einstein sought to describe gravitation in a way that was independent of the motion of observers and of the coordinates chosen to label events. His work led to a geometrical theory that described gravity purely by the structure of the space-time continuum. According to this geometrical theory, gravity affects all forms of matter and energy, all of which move in space-time. Thus Einstein's theory obeys the weak equivalence principle and gives the same gravitational acceleration for all freely falling objects.

In addition to describing the effect of gravity on matter, Einstein described the effect of matter on gravity. This theory, which Einstein completed in 1915, is called general relativity. Although Einstein's theory is much different from Newton's, it predicts nearly the same effects in systems in which gravitational fields are weak and velocities are slow compared to the velocity of light. Planetary motion had provided a particularly accurate verification of Newton's theory, but Einstein's accounted for some phenomena in the solar system not considered by Newton.

One such phenomenon was the perihelion precession of Mercury. In the 19th century, it was observed that the rate of precession differed by 43" per century from what Newton's theory predicted; Einstein's theory predicted precisely such a difference in the precession rate. Another such phenomenon is the bending of light rays by the Sun's gravitational field, which Newton's theory did not predict at all. Einstein's prediction was confirmed by Arthur S. Eddington during a total eclipse in 1919, and later by others to 1% accuracy. Einstein also predicted the gravitational red shift, a change in the frequency of electromagnetic waves escaping from a strong gravitational field, which was confirmed by Robert Pound and Glen Rebka in 1960. In 1964, Irving Shapiro used general relativity to predict a time delay of signals passing near the Sun, an effect since confirmed.

In addition, Einstein's theory of general relativity predicts several qualitatively new effects in other systems and is especially useful in dealing with cosmology. Relativity asserts that the universe must be either expanding or contracting. Einstein was not sufficiently bold to believe this prediction, however, so he modified his equations to allow for a static universe. In 1929, though, Edwin Hubble discovered that the universe is expanding. Whether gravity will eventually cause it to collapse is a subject of current investigation and debate. General relativity also predicts gravitational waves from masses in nonuniform motion, but these waves are so weak that they have not yet been definitely detected. Finally, Einstein's theory predicts gravitational collapse of sufficiently massive objects into black holes. Today there is mounting evidence that several astronomical systems may contain black holes.

Einstein's general relativity is not the only 20th-century theory of gravity, though it is perhaps the simplest and most elegant. All viable theories of gravity must, like Einstein's, be complete, self-consistent, and relativistic. They must also have the correct Newtonian limit, uphold the weak equivalence principle, and predict the same Einstein shift as measured by all ideal clocks at the same position. There is strong experimental evidence for accepting these criteria as fundamental, and it has been conjectured by L. I. Schiff that they can be satisfied only by geometrical, or metric, theories.

The strongest rival theory has been the Brans-Dicke theory. Like general relativity, it is a geometrical theory that satisfies the fundamental criteria. Its field equations are different, however, and it claims that the geometry of space-time is affected not only by matter but also by an additional scalar field. Unlike Einstein's calculations, the Brans-Dicke theory cannot predict the perihelion shift of Mercury.

Some recent theories attempt to explain gravitation non-geometrically, proposing instead that particles called gravitons are responsible. These so-called supersymmetry theories place gravitational phenomena within the realm of quantum physics. They are part of an attempt to show that the four fundamental interactions of nature are related, and that they were a single, united force at the birth of the universe.

Multidimensional analysis has also raised the issue of the constancy of the gravitational constant G. This idea was proposed earlier by British physicist Paul Dirac in his so-called "large numbers" hypothesis. Dirac noted that the ratio of the strength of the electromagnetic force to that of the gravitational force (approximately 10(40)) is roughly equivalent to the age of the universe in atomic terms. He wondered whether there might be a deep physical connection requiring this similarity and proposed that this would be the case if G slowly decreased as the age of the universe increased. If G were decreasing, however, gravitational time would change with respect to atomic time; thus far, experiments have not shown this to be happening.

Unified Field Theory

In theoretical physics, a unified theory may be defined most broadly as a theory that with one set of equations would unify all four of the fundamental interactions of nature: gravitation, electromagnetism, and the strong and weak nuclear forces. No such all-encompassing theory yet exists.

Historically, the term unified field theory was first given to the attempts by Albert Einstein to unify two of the fundamental forces, gravitation and electromagnetism. That is, after presenting his theory of gravitation—the general theory of relativity—he set out to reconcile his results with the description of electromagnetism given by the famous set of equations of James Clerk Maxwell. Einstein's idea was to combine his and Maxwell's descriptions by treating both subjects from an essentially geometric point of view. He failed to achieve this goal. This failure, however, must be judged in light of the fact that at that time the strong and weak nuclear forces had not yet been discovered.

Other attempts to incorporate electromagnetism into the basically geometric formalism of general relativity were made by the German physicist Hermann Weyl and the American John Wheeler. Although some of these theories are more aesthetically pleasing than others, all lack the connection with quantum phenomena that is so important for interactions other than gravitation.

Later attempts at unification have been made from the quite different point of view of merging the quantum field theories that describe, or are supposed to describe, the four fundamental interactions. The most successful thus far has been the electroweak theory of Sheldon Glashow, Steven Weinberg, and Abdus Salam, which joins electromagnetism and the weak interaction. In the simplest version of this theory, forces are transmitted by the exchange of four different types of particles called bosons, which are assumed to be massless. By means of a "broken symmetry" an effective generation of masses occurs, with three bosons, W(+), W(-), and Z(0) particles, having masses on the order of 50 to 100 times the mass of the proton, and a 4th boson, the photon, remaining massless. The W and Z bosons were detected by high-energy experiments at the CERN laboratories in 1983. Weinberg, Salam, and Glashow shared the 1979 Nobel Prize for physics for their model.

Many other unified theories have been proposed, beyond the electroweak theory. Some try to involve the strong interaction, and some "theories of everything" try to include gravitation as well. The latter are also known as supersymmetry theories. None have been successful as yet.

Quantum Mechanics

According to Einstein's general theory of relativity, nothing can travel at the speed of light or greater speeds. However, quantum mechanical theory allows for a phenomenon called tunneling that, in theory, would provide a method of moving one place to another at speeds greater than light. William Corliss mentions that in 1995, German researchers claimed to have transmitted Mozart's 40th Symphony a short 5 inches, but at 4.7 time the speed of light!

Quantum mechanics is the fundamental theory used by 20th-century physicists to describe atomic and subatomic phenomena. It has been successful in tying together a wide range of observations into a coherent picture of the universe.

While quantum mechanics uses some of the concepts of Newtonian mechanics, the previous description of physical phenomena, it differs fundamentally from the Newtonian. For example, in Newtonian physics, quantities were believed to be continuously variable, able to take on any value in some range. An example is angular momentum, which, for a particle revolving in a circular orbit about some center of attraction, is proportional to the speed multiplied by the distance from the center. Because that distance could have any value in Newtonian mechanics, so could the angular momentum. In quantum mechanics, on the other hand, angular momentum is always restricted to certain discrete values, whose ratios are simple rational numbers.

An even more fundamental difference between quantum mechanics and previous physical theories is that probability enters in a basic way into how quantum mechanics describes the world. This is made evident by the ways that quantum mechanics and Newtonian mechanics deal with predictions of the future. For something described by Newtonian mechanics, such as the solar system, it is possible, if sufficiently accurate measurements are made at one time, to predict the future behavior of the system to arbitrarily great accuracy. For systems described by quantum mechanics, even one as simple as an atom with a single electron, precise prediction of future behavior is usually impossible. Instead, only predictions of the probability of various behaviors can be made. This can be illustrated by the description of an unstable radioactive nucleus. Quantum mechanics does not predict when the individual nucleus will decay, although if many similar nuclei are surveyed, one can predict what fraction will decay in any time interval. This novel feature of quantum mechanics, known as indeterminism, has been one of the things that has led some prominent physicists, such as Albert Einstein, to resist it. Nevertheless, it appears to be an unavoidable feature of physics at the atomic and subatomic levels.

Early Historical Development: Planck's Work

Quantum mechanics was developed over a period of 30 years, during which it was successively applied to several physical phenomena. The first use of quantum ideas was made in the analysis of how electromagnetic radiation is produced. This was done by the German physicist Max Planck in 1900. Planck was trying to account for the distribution, among different frequencies, of the radiation emitted by a hot object, such as the surface of the Sun. He found that to obtain results in agreement with observation, he had to assume that the radiation was not emitted continuously, as was previously believed. Instead, it was emitted in discrete amounts, which he called quanta. For these quanta, there was always a relation between the frequency f, and the amount of energy emitted E, of the form $E=hf$. Here h is a universal constant introduced by Planck, and now named after him. Planck's constant has the units of energy multiplied by time, known as action. Its numerical value is approximately 6.63 x 10(-34) joule-seconds. The specific result of Planck's analysis was a formula expressing the amount of energy radiated at any frequency as a function of the temperature of the emitting object. This relation, the blackbody distribution, agrees accurately with observation.

In Planck's work, the nature of the quanta was rather mysterious. It was clarified by the work of Einstein, who in 1905 proposed that light itself was composed of individual packages of energy, which later came to be known as photons. Einstein also proposed that the frequency of the light is related to the energy of the photons composing it by Planck's formula. Einstein's theory of light quanta, which was rejected by many of his contemporaries, including Planck, was verified both by Robert Millikan's work on the photoelectric effect, and by the discovery by Arthur Compton of the Compton effect, or the scattering of photons by electrons.

Another significant early use of quantum ideas was by Niels Bohr, who in 1913 showed that by assuming that the angular momentum of electrons in a hydrogen atom could only take on values that are an integer multiple of Planck's constant divided by 2∏(pi), he could derive accurate expressions for the frequencies of light emitted by the atom. Bohr's analysis implied that only certain energy values are possible for the electron in the atom, that there is a minimum value, and that in this minimum energy state, the electron cannot radiate energy. This result helped explain how the atom could be stable, and how all atoms of one element have the same chemical properties. However, it proved impossible to extend Bohr's ideas directly to atoms more complex than hydrogen. Also, the strange blend of Newtonian and quantum ideas left physicists uneasy about the supposedly basic principles of their science.

Forms of Quantum Mechanics

The development of actual quantum mechanics—the mathematical theory—took place in the years 1924-27. Initially there were two seemingly different approaches: matrix mechanics, invented by Werner Heisenberg, and wave mechanics, invented by Erwin Schrodinger. However, it was soon shown that these were distinct aspects of a single theory, which came to be known as quantum mechanics. This unified version was invented by Paul Dirac. In matrix mechanics, physical quantities such as the position of a particle, are represented not by numbers, but by mathematical quantities known as matrices. Matrix mechanics is most useful in dealing with situations in which there is a small number of relevant energy levels, such as a particle with definite angular momentum in a magnetic field.

Wave mechanics is more useful in a situation where the number of energy levels is infinite, as with an electron in an atom. It is based on the idea originally suggested by Louis deBroglie, that particles such as electrons have waves associated with them. The wavelength, gamma, of the wave is related to the mass, m, and speed, v, of the particle by the relation gamma = h/mv. This implies that for electrons moving at 10 percent the speed of light, such as those produced by some television tubes, the wavelengths are about 10 (-10) meters, or about the distance between atoms in a crystalline solid. This prediction of deBroglie was verified by Clinton Davisson and by George Thomson, who were able to pass the electron waves through metallic crystals, and so produce diffraction patterns similar to those produced by X rays.

In 1925, Erwin Schrodinger developed an equation, now bearing his name, which describes how a wave associated with an electron or other subatomic particle varies in space and time as the particle moves under the influence of various forces. This equation has many types of solutions, and Schrodinger imposed the condition that for a particle bound in an atom, the solution should be mathematically well defined everywhere. When applied to the case of an electron in a hydrogen atom, Schrodinger's equation immediately gave the correct energy levels previously calculated by Bohr. However, the equation could also be applied to more complicated atoms, and even to particles not bound in atoms at all. It was soon found that in every case, Schrodinger's equation gave a correct description of a particle's behavior, provided that the particle was not moving at a speed near that of light.

In spite of this success, the meaning of the waves remained unclear. Schrodinger believed that the intensity of the wave at a point in space represented the "amount" of the electron that was present at that point. That is, the electron was spread out, not concentrated at a point. However, it was soon found that this interpretation was untenable, because even if a particle was originally concentrated on a small region, in most cases it would soon spread over an increasingly larger region, in contradiction to what was the observed behavior of particles.

The correct interpretation of the waves was discovered by Max Born. While studying how quantum mechanics describes collisions between particles, he realized that the intensity of the deBroglie-Schrodinger wave was a measure of the probability of finding the particle at each point in space. In other words, a measurement would always find a whole particle, rather than a fraction of one, but in regions where the wave intensity was low, the particle would rarely be found, whereas in regions of high intensity, the particle would often be found.

Heisenberg's Uncertainty Relation

An important contribution to the interpretation of quantum mechanics was given in 1927 by Heisenberg. He analyzed various "thought experiments" that were designed to suggest information about the location and velocity of a particle. An example would be the use of a microscope to image an electron. It is known that, because of the wave properties of light, a precise electron image requires the use of light of very short wavelength, and therefore high frequency. However, the Planck- Einstein relation implies that for such light, the photons must carry a large amount of energy and momentum. In the collisions between such photons and electrons, the electron momentum will be changed uncontrollably from what it was before the collision. As a result, the increased precision with which the electron's position is known is unavoidably accompanied by a loss of accuracy in the knowledge of its momentum. On the basis of this and related analyses, Heisenberg was led to formulate his uncertainty principle, which in its simplest form, states a reciprocal relation between the uncertainty delta x, with which we can know the position of any object, and simultaneously, the uncertainty delta p, with which we can know its momentum. The mathematical statement of the uncertainty relation is given by (delta x) (delta p) is less than h/4(pi). For an object of everyday size this limitation on simultaneous measurements is very unimportant, when compared to ordinary experimental uncertainties. For this reason, there is rarely any significant difference between the predictions of Newtonian and quantum mechanics for such objects. However, for an electron in an atom, the uncertainty restrictions are so significant that they essentially determine both the size and the minimum energy of the atom.

With Born's probability interpretation of the wave intensity and Heisenberg's uncertainty principle, the elements of the standard indeterministic interpretation of quantum mechanics were in place by 1930. This interpretation is often known as the Copenhagen interpretation, because Niels Bohr, who made important contributions to its formulation, ran an influential physics institute there during this period. However, many physicists, including Einstein and Schrodinger, who accepted the mathematical formulation of quantum mechanics, were uncomfortable with the Copenhagen interpretation, and criticized it. The question of the correct interpretation of the mathematical formalism has remained something of a problem.

Directly after its discovery, quantum mechanics was applied to many problems in atomic physics and chemistry, such as the structure of many-electron atoms and of molecules. These applications were generally successful in explaining old observations and in predicting newer ones. An example of the latter case was the successful prediction that hydrogen molecules could exist in two types, depending on the relative orientation of the angular momentum of the nucleus. This type of success led Paul Dirac, in 1928, to describe quantum mechanics as "including all of chemistry and most of physics." Although the second half of this statement has not proved to be perfectly accurate, extensions of quantum mechanics have been successful in explaining an ever-growing number of physical phenomena. For example, in the 1930s and 1940s, George Gamow used quantum mechanics to explain radioactive alpha decay of atomic nuclei.

For some applications to atomic nuclei, and for accurate calculations in atomic physics, it became necessary to extend the original form of quantum mechanics to make it consistent with Einstein's special theory of relativity. This was first done by Dirac in 1927, with an equation bearing his name. Dirac's equation proved immediately successful in accounting for a property of electrons known as spin. Spin is angular momentum of rotation about an axis through the electron, somewhat like that of the Earth rotating about its own axis. It was previously known that all electrons carry a spin of h/4(pi), but the reason was not clear. The Dirac equation explained this and accurately accounted for some magnetic properties of spinning electrons. It also made a novel prediction of the existence of particles similar in mass and spin to electrons, but with opposite electric charge. These particles, which have come to be known as positrons, were discovered by Carl Anderson in 1932. They were the first example of antiparticles, whose existence is predicted by any theory that satisfies the requirements both of quantum mechanics and special relativity.

Quantum Field Theory

The study of antiparticles and their properties highlighted a new aspect of relativistic quantum theories—the creation and annihilation of matter. Dirac had predicted, and it was soon observed, that electrons and positrons could be created together in pairs, when high-energy photons passed through matter. Furthermore, a positron that comes near to an electron quickly disappears together with the electron, converting into several photons. In order to describe transformations in which the number of particles changes, it was necessary to apply quantum mechanics to a new area—that of fields.

In Newtonian physics, a field represents a physical quantity, such as electric force, which varies from point to point in space and time according to precise mathematical equations. Such classical fields can have any numerical value at any point. The general version of quantum theory was first applied to the electromagnetic field by Dirac, who showed that this combination automatically implied the existence of photons with the properties assumed by Planck and Einstein. Furthermore, Dirac was able to use this quantum field theory formalism, which came to be known as QED, or quantum electrodynamics, to describe how photons are emitted and absorbed by charged particles, as when an electron in an atom radiates. An important practical application of QED in the late 1950s was the invention of the laser.

A number of physicists applied similar ideas to other, previously unknown fields in order to describe processes in which the numbers of other types of particles change. For example, Enrico Fermi, in 1933, used quantum field theory to explain the emission of electrons from a nucleus, the process known as beta decay. The general lesson learned from this is that fields satisfying the laws of quantum mechanics and relativity automatically describe particles that can be created or destroyed.

Quantum field theory had some unforeseen consequences. One aspect of Heisenberg's uncertainty principle is that the law of conservation of energy is not strictly observed for short periods of time. Because of this, a particle such as an electron can briefly emit and then reabsorb other particles, such as photons. These transients, called virtual particles, influence the properties that we measure for the electron. In particular, they change its mass from what it would have been if they did not exist. The extra mass due to virtual particles is called the self mass. Unfortunately, when physicists in the 1930s tried to calculate the self mass due to virtual photons, they got an infinite result. For some time, this result paralyzed progress in quantum field theory. However, in the 1940s, a method was found for dealing with infinite self mass and certain related infinities. This procedure, known as renormalization, has dominated quantum field theory since that time.

Renormalization

The idea behind renormalization is that the self mass is not directly measurable. Only the combination of self mass and any intrinsic mass that the electron might have can be observed. It was suggested, first by Hendrik Kramers, that an infinite self mass might combine with an infinite intrinsic mass to give the finite observed mass. It should then be possible to express all other observable quantities in terms of this sum, avoiding the problem of infinities. Calculations involving this procedure, known as mass renormalization, are quite delicate to carry out. Indeed, they were only done successfully after new techniques were introduced in the late 1940s by Julian Schwinger and Richard Feynman. These methods are designed to be consistent with relativity theory at all stages, unlike earlier methods, which made sharp distinctions between space and time. Feynman's methods involve the use of suggestive pictures, now called Feynman diagrams, which are correlated with any process to be calculated. For example, the emission of a photon by an electron is pictured as a solid line of indefinite length, representing the electron, with a wavy line, representing the photon, originating in the middle of the electron line. Feynman described a set of rules by which the probability of occurrence of any process could be calculated directly from the associated diagram.

In the late 1940s, using the methods of Feynman and Schwinger, scientists calculated small corrections, due to emission and absorption of virtual photons, for the energies of electrons in hydrogen atoms, and for the magnetic properties of electrons. These calculations, which have continued to ever higher levels of accuracy, in some cases agree with

observation to the incredible accuracy of one part in a billio
probably the greatest triumph that theoretical physics has yet achi

The success of QED led many physicists to believe th renormalizable quantum field theories could be found to describe properties of subatomic particles that are not included in QED, such as the strong forces that bind neutrons and protons into nuclei, and the weak forces responsible for beta decay. For many years this hope was not realized, because not enough was known either about types of renormalizable theories or about the particles to which such theories should be applied. This situation changed in the 1960s and 1970s, following the invention, by Chen-Ning Yang and Robert Mills, of a particular kind of renormalizable quantum field theory known as a gauge field theory. It was shown that one type of gauge field theory, named quantum chromodynamics, or QCD, was capable of describing the strong interactions provided that it was applied not to protons and neutrons, but to quarks, hypothetical particles that compose protons and neutrons and other particles affected by strong interactions. A second gauge field theory was shown by Sheldon Glashow, Steven Weinberg, and Abdus Salam, to be capable of describing electromagnetic and weak interactions together, thus unifying in a single theory two important aspects of nature.

In spite of the success of renormalizable quantum field theories, some prominent theoretical physicists, such as Dirac, have expressed misgivings about them. Although observable quantities are finite in these theories, this is achieved through manipulations with infinite quantities that are mathematically suspect and aesthetically unpleasant.

The Interpretation of Quantum Mechanics: Hidden Variable Theories

Quantum mechanics is now over 60 years old and has been very successful in providing explanations for physical phenomena. Nevertheless, there remains a dissatisfaction among some physicists both with the theory itself and with the prevailing Copenhagen interpretation. Much of this criticism derives from the radical change from earlier theories that quantum mechanics represents. Some of it involves problems that arise within quantum mechanics itself.

One criticism of quantum mechanics relates to its indeterminacy. This was Einstein's original objection, although he later developed others. Because an individual radioactive nucleus will eventually decay at a specific time, Einstein and others believed that a complete physical theory should allow this time to be predicted exactly, rather than just statistically. Einstein did not specify what type of theory he had in mind to replace quantum mechanics, but others have suggested that the solution be sought in some type of "hidden variable" theory. In hidden variable theories, physical properties other than those we can yet measure would determine those events about which quantum mechanics can only make probability predictions. The mathematician John von Neumann proved long ago that no hidden variable theory can agree exactly with the predictions of quantum mechanics. The predictions have not all been examined, however, so there exists some possibility a hidden variable theory could be formulated that agrees with all observations that have been made. None has yet been produced that physicists find satisfactory.

A second problem with the interpretation of quantum mechanics, which troubles even those who accept the theory, involves the idea of measurement. Schrodinger's wave equation can be used to describe how any system changes from one time to another. If the wave is known everywhere in space at one time, it can be predicted everywhere at a later time. However, knowing the wave intensity only allows for probability predictions of the results of measurements. When a measurement is actually made, the observer suddenly obtains exact information about at least one property of the system, such as its energy. This change from probabilistic to exact information has come to be called the reduction of the wave function. It has been proved that even if the interaction between the measuring instrument and the system being observed is taken into account, this reduction cannot be properly described by the Schrodinger equation. Various scientists have taken different attitudes toward this result. Some have championed the view that the consciousness of an observer plays a fundamental role in reduction of the wave function. Others have argued that because quantum mechanics cannot account for reduction, it is incomplete. Perhaps the most widely accepted view is that reduction of the wave function always involves the interaction of a microscopic object, such as an electron, with a macroscopic system, the measuring instrument. When this interaction takes place, there is an irreversible change in the measuring instrument, and it is this change that results in the reduction of the electron's wave function. There may be some truth to this view but it only solves part of the problem, because irreversibility itself is not completely understood.

The Einstein-Podolsky-Rosen Paradox and Bell's Theorem

Another problem concerning the interpretation of quantum mechanics derives from work by Einstein with Boris Podolsky and Nathan Rosen. In an article published in 1935, they pointed out that the predictions of quantum mechanics—in particular, the idea of indeterminacy—in some cases were in conflict with what they considered to be a plausible criterion for reality. The situation envisaged was one in which a physical system that is originally a whole splits into two parts that eventually become widely separated. The reality condition they imposed was that measurements done on one part of the separated system should not affect the other part of the system. An example of their analysis is an atom containing an electron and proton, each with spin angular momentum of h/4(pi), but with total spin zero, which breaks apart so that the electron goes in one direction and the proton in another. Because of the conservation of angular momentum, the total angular momentum of the electron and proton remains zero even when they are widely separated. If a physicist determines that the electron spin lies along some direction, he can immediately infer that the proton spin points in the opposite direction. The reality condition would then imply that the proton spin was already determined to have this value before the measurement was made on the electron.

An argument given by John Bell, in 1964, dealt with the fact that Einstein's conclusion contradicts quantum mechanics. His finding, known as Bell's theorem, derives from statistical measurements of spin values of many correlated electrons and protons. It states that any theory satisfying Einstein's reality condition—that reality is a localized phenomenon, and particles have determined properties—necessarily implies a relation among the results of a series of measurements. Aseries of experiments to test Einstein's reality condition, and Bell's theorem, have been carried out. The results do not support Einstein's reality condition, but instead support quantum mechanics.

Research that employs quantum mechanics remains at the center of contemporary physics. One aspect of this research involves the search for approximate methods that can be applied with the basic principles of quantum mechanics in studies of situations that are so complex that they cannot be dealt with exactly. Much of the research in condensed-matter physics is of this nature. An important discovery in this area is that in some situations the discreteness of physical quantities that usually occurs on the subatomic level can also occur on the macroscopic level. The quantized Hall effect, a property of electrical resistance of certain substances under the influence of electric and magnetic forces, is a recently discovered example of this.

Another important area of research involves the attempt to include gravity among the phenomena that can be described by quantum mechanics. Although there are no observations yet that require the use of a quantum theory of gravity, physicists believe such phenomena may occur inside black holes and may have occurred everywhere in the earliest moments of the universe. It has not yet been possible to formulate a consistent quantum theory of gravity, either by beginning with Einstein's nonquantum general theory of relativity or by applying the usual ideas of quantum field theory. Currently, some physicists are pursuing an approach based on a quantum theory of strings, objects extended in one spatial dimension, as opposed to conventional particles, which are extensionless points. String theories may succeed in uniting gravity with the other forces of nature in a unified quantum-mechanical description of nature. Meanwhile, there is no doubt that quantum mechanics is the most successful theory of physical phenomena yet invented by the human mind.

Superstring Theories and Grand Unification Theories

In theoretical physics, superstring theories are mathematical models that describe fundamental particles as extremely short (about 10(-35) m), one-dimensional strings, rather than as zero-dimensional points. It is imagined that these strings are capable of twisting, looping, combining, and separating. They are also assumed to exist in more than four dimensions, with the additional dimensions rolled up in a tight space. Superstring theories are extremely speculative and remain unproved, but certain successful predictions of the models offer hope that they might eventually lead to a single theory of forces and particles. The earliest string theories have been used, unsuccessfully, in attempts to understand the strong nuclear force. They have since been combined with theories of supersymmetry, hence the name superstring.

In theoretical physics, grand unification theories (GUTs) are attempts to describe three of the fundamental forces of nature—the strong, weak, and electromagnetic forces—as aspects of a single interaction. Grand unification theories also describe the two primary constituents of matter, quarks and leptons, as manifestations of a single type of subatomic particle.

According to current theory, forces between subatomic particles are transmitted via the emission and absorption of vector bosons, particles with one unit of intrinsic angular momentum. Theorists use a mathematical structure known as a gauge field to describe these fundamental interactions. The weak, strong, and electromagnetic interactions differ mainly in their strengths, as measured by the rest energies of their bosons, and by the probability of boson emissions. In order of decreasing strength, the individual bosons are gluons (the strong force), photons (the electromagnetic force), and W and Z particles (the weak force).

In GUTs the differences between fundamental forces are seen to be the result of the fact that particles are observed at relatively low energies. If particles could be observed at very high energies, the three types of interaction would be found to have equal strength. In addition, quarks and leptons would behave similarly under those conditions. The energy at which this unification is expected is about 10(14) power GeV (giga electron volts), a trillion times higher than the energies of present-day experiments. While it may never be possible to examine such high-energy processes in the laboratory, it is thought that in the very early universe, a small fraction of a second after the Big Bang the average energy of all particles was high enough for these particles to behave in the manner described by GUTs.

One startling prediction of GUTs is that new interactions may exist, involving vector bosons whose rest energy is 10(14) power GeV. These interactions would allow three quarks to convert into a lepton, including the decay of protons and neutrons into leptons. The result of such decays would be a proton lifetime of 10(31) power years, and evidence for this would defeat the long-held notion of the eternal stability of the proton. Experiments designed to detect these rare proton decays are being carried out, but no conclusive evidence has been obtained.

Although the validity of GUTs is not yet established, physicists are developing theories that will link the fourth force, gravitation (and its boson, the graviton), with the other three forces. These models are referred to as supersymmetry theories.

Symmetry in Theoretical Physics

In physics, a system is said to exhibit a symmetry if it remains unchanged under a given operation. For example, a ball looks the same under rotation in any direction about its center. The ball is therefore said to have a spherical symmetry. Such symmetries play a fundamental role in the understanding of various physical phenomena. They are an especially important part in the study of elementary-particle physics, where the exact nature of the force laws is still unknown. The significance of symmetry is that many aspects of the behavior of a system may be predicted on the basis of its symmetry without a detailed knowledge of its inner workings.

An inventor of a time machine demonstrates it by sending the family cat back in time. An illustration from Ray Palmer's story "The Time Tragedy" for *Wonder Stories* (Dec. 1934). Palmer went on to become the editor of *Amazing Stories,* a magazine which began the Shaver Mystery and eventually lead to *Fate* magazine, which told "true" stories of time travellers, UFOs, and other supernatural phenomena.

A double-page illustration of the story "Thompson's Time Traveling Theory" by Malcolm Smith, published in Ray Palmer's *Amazing Stories* (1944). In the story a time-machine inventor makes an experimental test of the so-called grandfather paradox by going back in time and killing his grandfather.

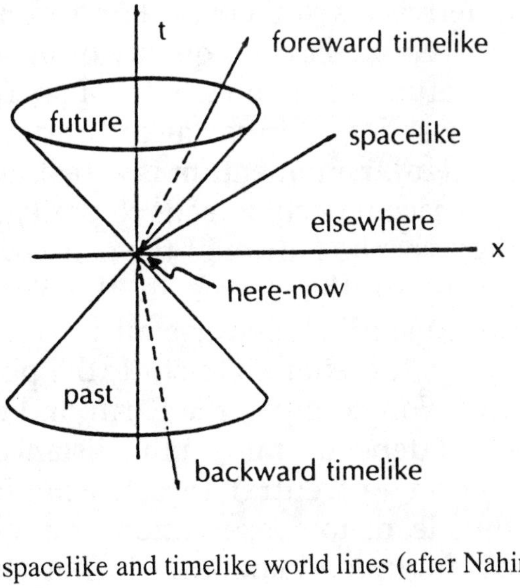

A light cone with spacelike and timelike world lines (after Nahin, 1993).

All symmetries may be classified into two types, discrete and continuous. For a discrete symmetry, there are only a finite number of operations that can lead to identical physical configurations: for example, a square drawn on a paper has a discrete symmetry because there are only four ways it can be rotated about an axis perpendicular to the plane of the paper to look unchanged. On the other hand, a circle drawn on a piece of paper looks unchanged after rotation through any angle about an axis perpendicular to the paper. Thus the circle exhibits a continuous rotational symmetry.

Some other examples of continuous symmetries are space-time translations and rotations. The former refers to all points in space and time being equivalent as far as physical laws are concerned. The latter refers to the equivalence of all directions in space (isotropy of space). Two important examples of discrete symmetry are space reflection and time reversal invariance. Many physical laws are unchanged under mirror reflection and under the reversal of the direction of time.

A conservation law states that some physical parameters of a system remain unchanged during the time evolution of the system. For example, in classical (nonrelativistic) physics, the total mass of a system is fixed before and after, for example, a collision process (the conservation law for mass). Similarly, the total linear and angular momentum and the total energy of a system obey conservation laws in both relativistic and nonrelativistic physics.

Since the advent of the Lagrangian formulation of physics, it has been established that the existence of conservation laws is related to the existence of underlying symmetries. For example, the conservation of linear momentum is a direct consequence of invariance of the Lagrangian under space translation, or homogeneity of space. Similarly, conservation of energy follows from invariance under time translation. The conservation of angular momentum is a result of rotational invariance. Symmetry under space reflection leads to parity invariance.

In order to discover the nature of the conservation laws operating in a system, the nature of the forces in the system and the associated symmetries must be studied. Because all forces derive their origins from the potential energy, it is enough to study the potential energy to discover conservation laws. For example, the Earth and the Sun attract each other through a force that depends only on the distance between the two and not on their orientation (a central force). The force law is rotationally invariant and thus leads to conservation of angular momentum. Because the direction of angular momentum is perpendicular to the plane of rotation, the Earth and Sun are locked in the same plane forever. Thus any two bodies interacting through a central force will have this property.

Symmetries allow only certain final states once the initial state is given, implying, in turn, that appearance of many final states is forbidden, or at least highly improbable. The rules governing such transitions are called selection rules and are important in the study of atomic systems, where knowledge of allowed initial and final states determines, for example, the intensity and energy of the emitted light of lasers.

Internal Symmetries and Space-Time

The symmetries discussed so far involve space or time, or both. Certain symmetries, though, are known to exist where the symmetry operation changes one type of particle into another. For example, it is known that in the nucleus the same force exists between a proton-proton, proton-neutron, and neutron-neutron pair, meaning that there is a substitution symmetry between the proton and the neutron. In an imaginary "space," where the proton and the neutron form two hypothetical directions, any rotation in that space will leave the nuclear forces unchanged. This feature is called isospin symmetry. The inclusion of additional particles in such a space leads to the higher unitary symmetries. Other examples of internal symmetry are electric charge, baryon number, and so on, which also owe their origin to operations in a fictitious internal space.

The internal symmetries may operate the same way or be different at each space-time point. In the latter case they are called gauge symmetries. In the 1960s theorists applied gauge symmetry to the concepts of both the weak and electromagnetic forces. They considered the idea of "spontaneous symmetry breaking," in which certain relations in theory can have exact symmetry, yet in their physical manifestations do not provide neat representations of the symmetry. The result was a single mathematical model that incorporated both forces: the electroweak theory.

What we have learned is that the fabric of space exists in different dimensions and time, as we know it, does not exist in the same way in all of these dimensions. It is theoretically possible to "tunnel" through time by accessing hyperspace. Incredibly, physics tells us that all things in the Universe are interconnected through the various fields—gravity, electricity, magnetism, the "weak" force of atomic structure, and time. In theory there is no barrier to time travel, rather, time travel and time anomalies are part of the fabric of the Universe itself!

The material in this chapter is largely from the *Grolier Multimedia Encyclopedia.*[39] Bibliographies given are as follows:

Gravity Bibliography: Bergmann, P. G., *The Riddle of Gravitation* (1968); Hawking, S. W., and Israel, W., *Three Hundred Years of Gravitation* (1987; repr. 1989); Mathews, P. M., et al., eds., *Gravitation, Quantum Fields and Superstrings* (1988); Misner, Charles, et al., *Gravitation* (1973); National Research Council, *Gravitation, Cosmology, and Cosmic-Ray Physics* (1986); Thorne, K. S., *Gravitational Radiation* (1989); Zee, Anthony, *An Old Man's Toy* (1989).

Relativity Bibliography: Aveni, A., *Empires of Time: Calendars, Clocks and Cultures* (1989); Coveney, P., and Highfield, R., *The Arrow of Time* (1991); Cowan, H. J., *Time and Its Measurement: From the Stone Age to the Nuclear Age* (1958); Davies, P., *About Time: Einstein's Unfinished Revolution* (1995); Elton, L. R. B., *Time and Man* (1978); Flood, R., and Lockwood, M., eds., *The Nature of Time* (1987); Fraser, J. T., *The Genesis and Evolution of Time* (1982); Howse, D., *Greenwich Time and the Discovery of the Longitude* (1980); Kopczynski, W., and Trautman, A., *Space, Time and Gravitation* (1992); Le Poidevin, Robin, and Macbeath, Murray, eds., *The Philosophy of Time* (1993); O'Malley, M., *Keeping Watch: A History of Time in America* (1990; repr. 1991); Toulmin, S. V., and Goodfield, J., *The Discovery of Time* (1976; repr. 1982); Whitrow, G. J., *The Natural Philosophy of Time,* 2d ed. (1981). Angel, R. B., *Relativity* (1980); Born, Max, *Einstein's Theory of Relativity,* rev. ed. (1962); Chaisson, Eric, *Relatively Speaking* (1990); Einstein, Albert, *The Meaning of Relativity,* 5th ed. (1956); Gardner, Martin, *Relativity for the Millions* (1962); Gibilisco, Stan, *Understanding Einstein's Theories of Relativity* (1991); Russell, Bertrand, *The ABC of Relativity,* 3d ed. (1969; repr. 1985); Tauber, G. E., *Relativity* (1988); Will, Clifford M., *Was Einstein Right?* (1988).

Quantum Mechanics Bibliography: Bergmann, Peter G., *Introduction to the Theory of Relativity* (1942; repr. 1976); De Vega, H. J., and Sanders, N., eds., *Field Theory, Quantum Gravity, and Strings* (1986); Einstein, Albert, *The Meaning of Relativity,* 5th ed. (1956); Hadlock, Charles, *Field Theory and Its Classical Problems* (1979); Mohapatra, R. N., *Unification and Supersymmetry* (1986); Tonnelat, Marie A., *Einstein's Theory of Unified Fields* (1966). Bernstein, Jeremy, *Quantum Profiles* (1991); Chester, Marvin, *Primer of Quantum Mechanics,* rev. ed. (1992); Feinberg, Gerald, *What Is the World Made Of?* (1977); Han, M. Y., *The Probable Universe* (1992); Healey, R. A., *The Philosophy of Quantum Mechanics* (1991) ; Herbert, Nick, *Quantum Reality* (1985); Jammer, Max, *The Conceptual*

Development of Quantum Mechanics (1966); Pagels, Heinz R., *The Cosmic Code* (1982); Wheeler, John, and Woyciech, Zurek, *Quantum Theory and Measurement* (1983).

Superstring Bibliography: Freund, P. G., and Mahanthappa, K. T., eds., *Superstrings* (1988); Gates, S., Jr., and Mohapatra, R. N., eds., *Superstrings* (1987); Lindley, David, *The End of Physics* (1993); Peat, F. D., *Superstrings and the Search for the Theory of Everything* (1989). Bernstein, Jeremy, *The Tenth Dimension* (1989); Feinberg, Gerald, *What is the World Made Of?* (1977); Fritzsche, Harald, *Quarks* (1983); Pagels, Heinz, *The Cosmic Code* (1982; repr. 1984) and *Perfect Symmetry* (1986); Weinberg, Steven, *Subatomic Particles* (1983; repr. 1990).

Symmetry Bibliography: Bloch, P., et al., eds., *Fundamental Symmetries* (1987); Boardman, Allan D., et al., *Symmetry and Its Applications in Science* (1973); Coxeter, H. S., *Introduction to Geometry,* 2d ed. (1989); Emmerson, J. M., *Symmetry Principles in Particle Physics* (1972); Gardner, Martin, *The Ambidextrous Universe: Mirror Asymmetry and Time-Reversed Worlds,* rev. ed. (1979); Lichtenberg, D. B., Unitary *Symmetry and Elementary Particles* (1978); Ludwig, W., and Falter, C., *Symmetries in Physics* (1988); Rosen, J., *A Symmetry Primer for Scientists* (1983); Weyl, Herman, *Symmetry* (1952); Yale, Paul B., *Geometry and Symmetry* (1968).

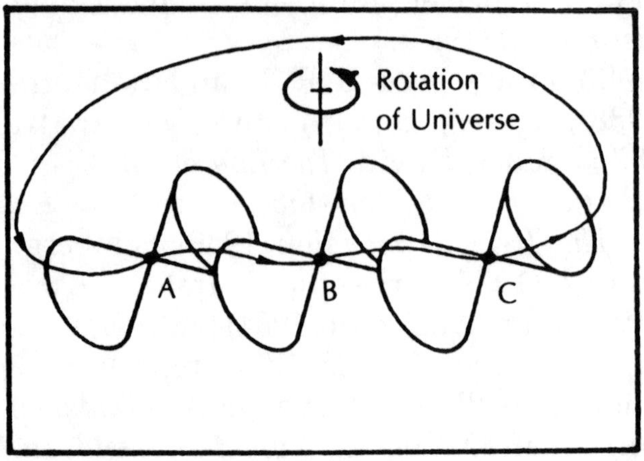

Tilted light cones in a rotating Universe (after Nahin, 1993).

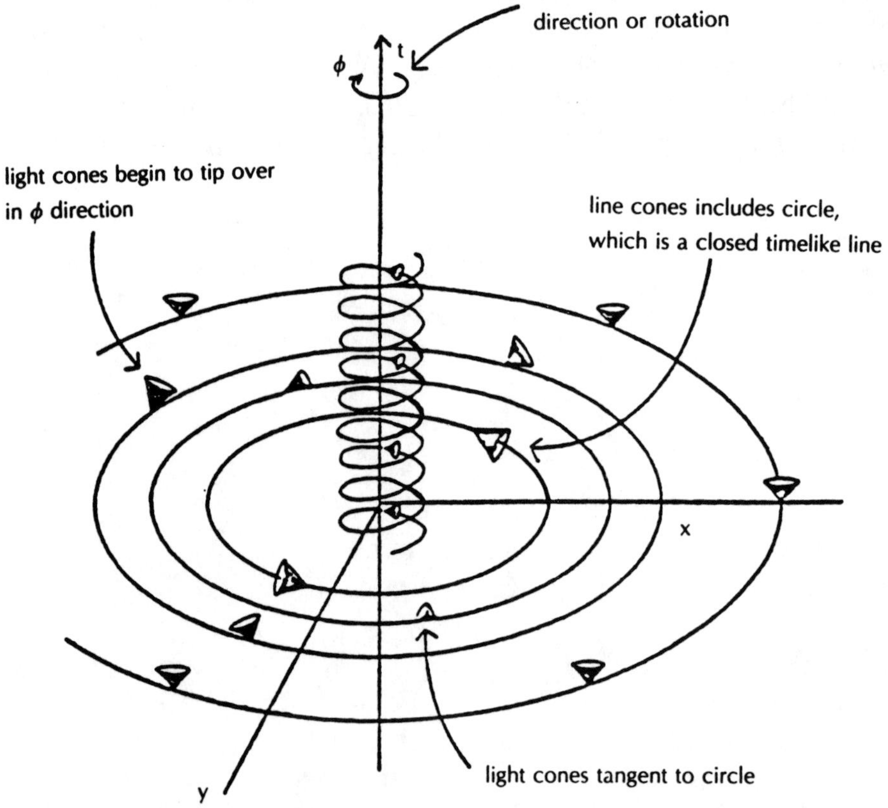

Future light cones point almost entirely in +t direction far from rotating matter; they begin to tip over as the matter is approached. Note that there is a helical timelike path that moves locally into the future in the -t direction: in other words, it goes into the past as seen by an observer far from the rotating matter. (The world lines of rotating matter are helixes in the +t direction, after Nahin, 1993).

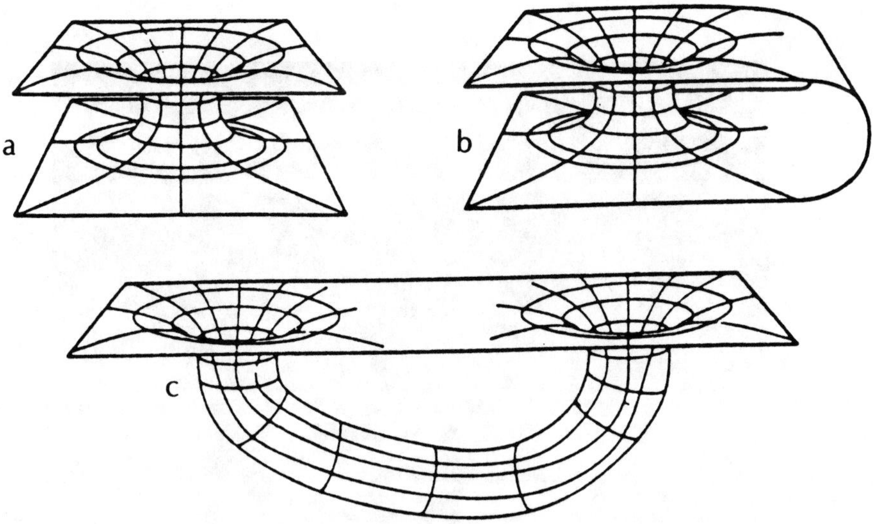

Time Tunnels and Wormholes. These are unavoidably misleading two dimensional interpretations of a time wormhole through three dimensional space. The wormhole in (a) connects two disjoint universes, while those in (b) and (c) are connections in the same universe. The wormhole "handle" can either be long or short compared to the distance in external space between the wormhole mouths.

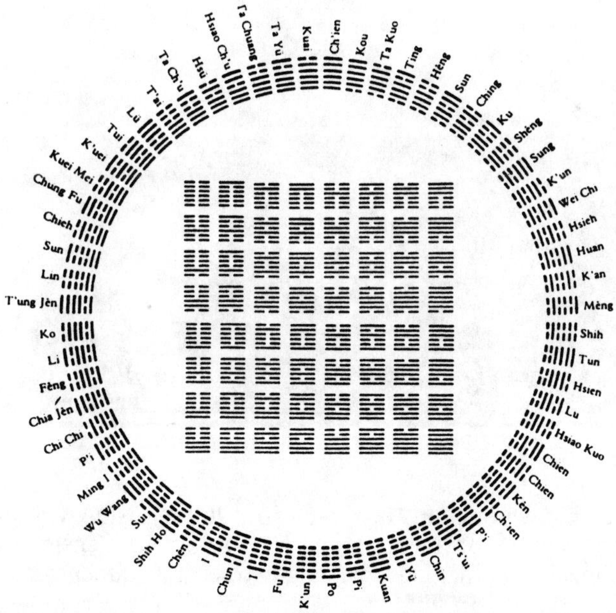

The 64 hexagrams of the *I Ching* displayed in the traditional arrangement of a circle and a square. This ancient oracle allows a type of time travel, as it can "predict" coming events, or allow the user to change the future, by acting as a two dimensional representation of parallel worlds.

A gigantic, cyclonic waterspout, observed in 1969 by NOAA aircraft near Key West, Florida. A good example of a vortex and how a three-dimensional media can suddenly become a one dimensional filament spanning two separate dimensions of space/time.

Table 1: A New Viewpoint

Quantum Effects Arise from a Matter, Zero-Point Energy Interaction

Quantum Event	Qualitative Explanation
Photon	Resonant absorption. Wave chopping occurs at detector
Quantum Eigenstates	Jump resonances of a nonlinear system
Ground State Stability	Zero-point radiation pressure balances Coulomb attraction.
Photoelectric Effect, Comptom Effect	See Scully and Sargent[6]
Blackbody Radiation	See Boyer[5]
Uncertainty Principle	Zero-point energy produces Brownian motion
Spontaneous Emission	Zero-point energy absorption
Pair Production	Soliton formation
Tunneling, EPR Paradox, Bell's Theorem, Nonlocal Connections[24,25]	Wheeler's "wormholes"[4] Hyperspace connections[27]
Infinite Self-Energies	Infinite zero-point energy flux implies higher dimensions of space
Renormalization	(Net energy of coherence) = (Infinite self-energy) - (Infinite incoherent zero-point energy)
Wave-Particle Duality	Waves are cohered zero-point energy; particles are solitons

From *Tapping the Zero Point Energy* by Moray B. King.

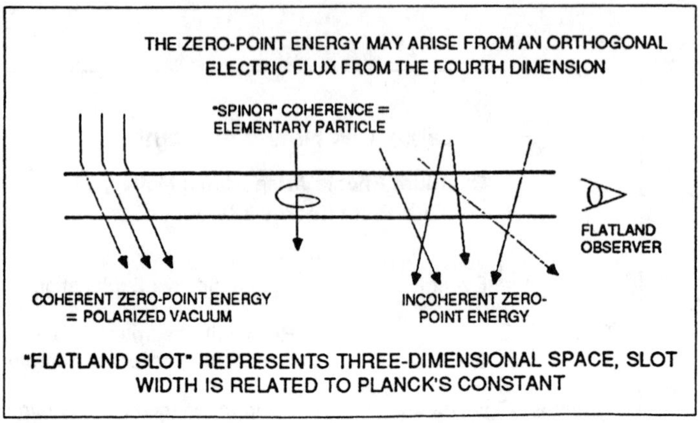

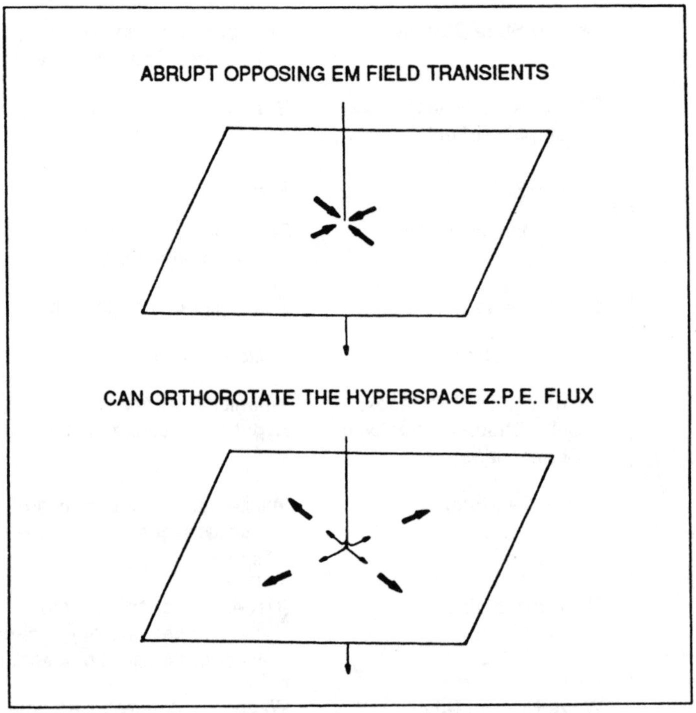

From *Tapping the Zero Point Energy* by Moray B. King.

2.

Relativity & Space-Time Curvature

From Entropy to Time Reversal Invariance and Beyond

In science one tries to tell people, in such a way as to be understood
by everyone, something that no one ever knew before.
But in poetry, it's the exact opposite.
—*Paul Dirac (1902-1984)*

Mathematicians are like Frenchmen:
whatever you say they translate into their own language,
and it means something very different.
—*Attributed to Goethe*

The desire for a mechanical time travel device takes us into the realms of relativity and space-time curvature. In the *Star Trek* television and movie series, the *USS Enterprise* would achieve "warp" speeds, where faster-than-light travel strained the dilithium crystals as the *Enterprise* sought to gain its destination as fast as possible.

However, in the *Star Wars* series by George Lucas, the spaceships are capable of jumping instantly from place to place, from solar system to solar system. Rather than running at "warp" speed, they are jumping through hyperspace. This is not faster than light, it is "other" than light. It is travelling by jumping from dimension to dimension and transcending time and space at the same time. No point whose coordinates can be calculated is too far away. It can be reached in an instant. It is time travel as well. To conquer hyperspace, an in-depth knowledge of relativity and space-time curvature is required.

From Entropy to Time Control

Entropy is the scientific term for the degree of randomness or disorder in processes and systems. In the physical sciences the concept of entropy is central to the descriptions of the thermodynamics, or heat-transfer properties, of molecules, heat engines, and even the universe as a whole. It is also useful in such diverse fields as communications theory and the social and life sciences.

Entropy was first defined by the German physicist Rudolf Clausius in 1865, based in part on earlier work by Sadi Carnot and Lord Kelvin. Clausius found that even for "perfect," or completely reversible, exchanges of heat energy between systems of matter, an inevitable loss of useful energy results. He called this loss an increase in entropy and defined the increase as the amount of heat transfer divided by the absolute temperature at which the process takes place. Because few real processes are truly reversible, actual entropy increases are even greater than this quantity. This principle is one of the basic laws of nature, known as the Second Law of Thermodynamics.

The First Law of Thermodynamics states that energy is conserved; no process may continuously release more energy than it takes in, or have an efficiency greater than 100%. The Second Law is even more restrictive, implying that all processes must operate at less than 100% efficiency due to the inevitable entropy rise from the rejection of waste heat. For example, large coal-fired electric power plants inevitably waste about 67% of the energy content of the coal. Other heat engines, such as the automobile engine and the human body, are even less efficient, wasting about 80% of available energy. An imaginary perpetual motion machine would have to defy these laws of nature in order to function. Such a machine, having its own output as its only energy source, would have to be 100% efficient to remain in operation. Friction always makes this impossible, for it converts some of the energy to waste heat.

The Arrow of Time

Another manifestation of entropy is the tendency of systems to move toward greater confusion and disorder as time passes. Natural processes move toward equilibrium and homogeneity rather than toward ordered states. For example, a cube of sugar dissolved in coffee does not naturally reassemble as a cube, and perfume molecules in the air do not naturally gather again into a perfume bottle. Similarly, chemical reactions are naturally favored in which the products contain a greater amount of disorder (entropy) than the reactants. An example is the combustion of a common fuel. Such reactions will not spontaneously reverse themselves. This tendency toward disorder gives a temporal direction—the "arrow of time"—to natural events.

A consequence of nature's continual entropy rise may be the eventual degrading of all useful energy in the universe. Physicists theorize that the universe might eventually reach a temperature equilibrium in which disorder is at a maximum and useful energy sources no longer exist to support life or even motion. This "heat death of the universe" would be possible only if the universe is physically bounded and is governed as a whole by the same laws of thermodynamics observed on earth.

The concept of entropy also plays an important part in the modern discipline of information theory, in which it denotes the tendency of communications to become confused by noise or static. The American mathematician Claude E. Shannon first used the term for this purpose in 1948. An example of this is the practice of photocopying materials. As such materials are repeatedly copied and recopied, their information is continually degraded until they become unintelligible. Whispered rumors undergo a similar garbling, which might be described as psychological entropy. Such degradation also occurs in telecommunications and recorded music. To reduce this entropy rise, the information may be digitally encoded as strings of zeros and ones, which are recognizable even under high "noise" levels, that is, in the presence of additional, unwanted signals.

The onset and evolutionary development of life and civilization on Earth appears to some observers to be in conflict with the Second Law's requirement that entropy can never decrease. Others respond that the Earth is not a closed system, because it receives useful energy from the Sun, and that the Second Law allows for local entropy decreases as long as these are offset by greater entropy gains elsewhere. For example, although entropy decreases inside an operating refrigerator, the waste heat rejected by the refrigerator causes an overall entropy rise in the kitchen. Life on Earth may represent a local entropy decrease in a universe where the total entropy always rises.

Ongoing work by the Belgian chemist Ilya Prigogine and others is aimed at broadening the scope of traditional thermodynamics to include living organisms and even social systems.

The experience of time, or of time duration, has received great attention in literature and philosophy. The experience varies among individuals and, because of its subjective nature, may appear to be inconsistent even in one individual. In scientific work, a numerical measure is used to order observations of events. If "now" is assigned the numerical value zero, then it is usual to assign earlier times negative values and later times positive values. To obtain a time scale, some periodic phenomenon that has repetitions occurring at a uniform rate that may be subdivided and counted must be used.

Before the 20th century it was assumed as self-evident that a single, universal, uniform time scale existed. For two events that are widely separated in space, it had been assumed that there is no difficulty in defining the meaning of the concept of simultaneity—namely, if to one observer the events appeared to occur simultaneously, then all other observers would agree that the events were indeed simultaneous. Albert Einstein, however, early in the 20th century, recognized that because of the universal constancy of the speed of light, the measurement of time depends on the motion of the observer.

Consider events A and B separated in space, which appear simultaneous to one observer; to another observer, then, who is in motion relative to the first, the event A may occur before or after B, depending on the direction of the relative motion between the two observers. Thus, in the modern view, time is no longer absolute, but dependent on the relative motion of observers making the time measurements. According to the theory of relativity, time is but one aspect of a more general four-dimensional space-time continuum, in which events occur in the universe. Time and space are different aspects of this underlying four-dimensional continuum. Frequently time is described as a "fourth dimension."

Time Scales

From earliest times, the rotation of the Earth (or the apparent location of the Sun in the sky) has been used to establish a uniform time scale. In order to specify a date, using the apparent motion of the Sun as a time scale, days must be counted from some reference date. In addition, a clock is used to measure fractions of a day.

Time derived from the apparent position of the Sun in the sky is called apparent solar time. Because of the eccentricity of the Earth's orbit around the Sun and the inclination of the Earth's rotation axis to the orbital plane, apparent solar time is not a uniform time scale. These effects can, however, be calculated and corrections applied to obtain a more uniform time scale called mean solar time. Universal Time (UT0) is equivalent to mean solar time at the Greenwich Meridian (Greenwich mean time, or GMT). Observations of the apparent motion of a distant star may be used to obtain yet another time scale used in astronomy, called sidereal time.

Additional small deviations from the uniformity of UT0 may be traced to small effects, such as the wandering of the Earth's polar axis and other periodic fluctuations of the Earth's rotation; accounting for these effects leads to additional, even more uniform, time scales (UT1 and UT2).

Ephemeris time is determined by the orbital motion of the Earth about the Sun and is not affected by fluctuations in the Earth's rotation. Astronomical observations may be used to determine ephemeris time to an accuracy of roughly 0.05 seconds, averaged over a nine-year period.

The invention of the quartz crystal oscillator and of the atomic clock makes possible the measurement of time and frequency more accurately than any other physical quantity. Thus, in addition to astronomical time scales, there are other time scales such as atomic time (AT), based on the microwave resonances of certain atoms in a magnetic field. Counting the cycles of an electromagnetic signal in resonance with cesium atoms provides an accuracy of a few billionths of a second over intervals of a minute or less.

Since about 1960 a number of laboratories around the world have cooperated in comparing their atomic time scales, leading to the formation of a weighted average of the various atomic time scales, which is now disseminated to the public as Universal Coordinated Time (UTC). In order to keep UTC in agreement with the length of the day, seconds are occasionally added to or deleted from the atomic time scale (a "leap second"). By international agreement, UTC is maintained within 0.7 seconds of the navigator's time scale, UT1.

The advancement of precision in time measurement has resulted in redefinitions of the second. Prior to 1956, one second was defined as the fraction 1/86,400 of the mean solar day. From 1956 to 1967, it was the ephemeris second, defined as the fraction 1/31556925.9747 of the tropical year at 00h 00m 00s 31 December 1899. The second is currently defined as the duration of 9,192,631,770 periods of the radiation corresponding to the transition between the two hyperfine levels of the ground state of the cesium-133 atom.

Relativity

Albert Einstein's theory of relativity has caused major revolutions in physics and astronomy during the 20th century. It introduced to science the concept of "relativity"—the notion that there is no absolute motion in the universe, only relative motion—thus superseding the 200-year-old theory of mechanics of Isaac Newton. Einstein showed that we reside not in the flat, Euclidean space and uniform, absolute time of everyday experience, but in another environment: curved space-time. The theory played a role in advances in physics that led to the nuclear era, with its potential for benefit as well as for destruction, and that made possible an understanding of the microworld of elementary particles and their interactions. It has also revolutionized our view of cosmology, with its predictions of apparently bizarre astronomical phenomena such as the big bang, neutron star, black hole, and gravitational waves.

The theory of relativity is a single, all-encompassing theory of space-time, gravitation, and mechanics. It is popularly viewed, however, as having two separate, independent theoretical parts—special relativity and general relativity. One reason for this division is that Einstein presented special relativity in 1905, while general relativity was not published in its final form until 1916. Another reason is the very different realms of applicability of the two parts of the theory: special relativity in the world of microscopic physics, general relativity in the world of astrophysics and cosmology.

A third reason is that physicists accepted and understood special relativity by the early 1920s. It quickly became a working tool for theorists and experimentalists in the then-burgeoning fields of atomic and nuclear physics and quantum mechanics. This rapid acceptance was not, however, the case for general relativity. The theory did not appear to have as much direct connection with experiment as the special theory; most of its applications were on astronomical scales, and it was apparently limited to adding minuscule corrections to the predictions of Newtonian gravitation theory; its cosmological impact would not be felt for another decade. In addition, the mathematics of the theory were thought to be extraordinarily difficult to comprehend. The British astronomer Sir Arthur Eddington, one of the first to fully understand the theory in detail, was once asked if it were true that only three people in the world understood general relativity. He is said to have replied, "Who is the third?"

This situation persisted for almost 40 years. General relativity was considered a respectable subject not for physicists, but for pure mathematicians and philosophers. Around 1960, however, a remarkable resurgence of interest in general relativity began that has made it an important and serious branch of physics and astronomy. This growth has its roots, first, beginning around 1960, in the application of new mathematical techniques to the study of general relativity that significantly streamlined calculations and that allowed the physically significant concepts to be isolated from the mathematical complexity, and second, in the discovery of exotic astronomical phenomena in which general relativity could play an important role, including quasars (1963), the 3-kelvin microwave background radiation (1965), pulsars (1967), and the possible discovery of black holes (1971). In addition, the rapid technological advances of the 1960s and '70s gave experimenters new high-precision tools to test whether general relativity was the correct theory of gravitation.

The distinction between special relativity and the curved space-time of general relativity is largely a matter of degree. Special relativity is actually an approximation to curved space-time that is valid in sufficiently small regions of space-time, much as the overall surface of an apple is curved even though a small region of the surface is approximately flat. Special relativity thus may be used whenever the scale of the phenomena being studied is small compared to the scale on which space-time curvature (gravitation) begins to be noticed. For most applications in atomic or nuclear physics, this approximation is so accurate that relativity can be assumed to be exact; in other words, gravity is assumed to be completely absent. From this point of view, special relativity and all its consequences may be "derived" from a single simple postulate. In the presence of gravity, however, the approximate nature of special relativity may manifest itself, so the principle of equivalence is invoked to determine how matter responds to curved space-time. Finally, to learn the extent that space-time is curved by the presence of matter, general relativity is applied.

Special Relativity

The two basic concepts of special relativity are the inertial frame and the principle of relativity. An inertial frame of reference is any region, such as a freely falling laboratory, in which all objects move in straight lines with uniform velocity. This region is free from gravitation and is called a Galilean system.

Free fall is motion determined solely by gravitational forces. For example, a dropped object, or one thrown into the air, is in free fall at every point in its trajectory. An object in space is always in free fall. Although influenced by the gravitational fields of many celestial bodies, it may not actually be "falling" toward any of them. A person in a freely falling vessel is also in free fall and experiences the phenomenon of weightlessness. Gravitational effects are the same on both person and vessel, so no acceleration is felt relative to the vessel.

The principle of relativity postulates that the result of any physical experiment performed inside a laboratory in an inertial frame is independent of the uniform velocity of the frame. In other words, the laws of physics must have the same form in every inertial frame. A corollary is that the speed of light must be the same in any inertial frame (because a speed-of-light measurement is a physical experiment) regardless of the speed of its source or that of the observer. Essentially all the laws and consequences of special relativity can be derived from these concepts.

The first important consequence is the relativity of simultaneity. Because any operational definition of simultaneous events at different locations involves the sending of light signals between them, then two events that are simultaneous in one inertial frame may not be simultaneous when viewed from a frame moving relative to the first. This conclusion helped abolish the Newtonian concept of an absolute, universal time. In some ways the most important consequences and confirmations of special relativity arise when it is merged with quantum mechanics, leading to many predictions in agreement with experiments, such as elementary particle spin, atomic fine structure, antimatter, and so on.

The mathematical foundations of special relativity were explored in 1908 by the German mathematician Hermann Minkowski, who developed the concept of the "four-dimensional space-time continuum," which includes time and the three spatial dimensions.

The Principle of Equivalence and Space-Time Curvature

The exact Minkowski space-time of special relativity is incompatible with the existence of gravity. Aframe chosen to be inertial for a particle far from the Earth where the gravitational field is negligible will not be inertial for a particle near the Earth. An approximate compatibility between the two, however, can be achieved through a remarkable property of gravitation called the weak equivalence principle (WEP): all modest-sized bodies fall in a given external gravitational field with the same acceleration regardless of their mass, composition, or structure. The principle's validity has been checked experimentally by Galileo, Newton, and Friedrich Bessel, and in the early 20th century by Baron Roland von Eotvos (after whom such experiments are named). If an observer were to ride in an elevator falling freely in a gravitational field, then all bodies inside the elevator, because they are falling at the same rate, would consequently move uniformly in straight lines as if gravity had vanished. Conversely, in an accelerated elevator in free space, bodies would fall with the same acceleration (because of their inertia), just as if there were a gravitational field.

Einstein's insight was to postulate that this "vanishing" of gravity in free-fall applied not only to mechanical motion but to all the laws of physics, such as electromagnetism. In any freely falling frame, therefore, the laws of physics should (at least locally) take on their special relativistic forms. This postulate is called the Einstein equivalence principle (EEP).

One consequence is the gravitational red shift, a shift in frequency f for a light ray that climbs through a height h in a gravitational field, given by (delta f)/f=gh/c(2) where g is the gravitational acceleration and c is the velocity of light. (If the light ray descends, it is blueshifted.) Equivalently, this effect can be viewed as a relative shift in the rates of identical clocks at two heights. A second consequence of EEP is that space-time must be curved. Although this is a highly technical issue, consider the example of two frames falling freely, but on opposite sides of the Earth. According to EEP, Minkowski space-time is valid locally in each frame; however, because the frames are accelerating toward each other, the two Minkowski space-times cannot be extended until they meet in an attempt to mesh them into one. In the presence of gravity, space-time is flat only locally but must be curved globally.

Any theory of gravity that fulfills EEP is called a "metric" theory (from the geometrical, curved-space-time view of gravity). Because the equivalence principle is a crucial foundation for this view, it has been well tested. Versions of the Eotvos experiment performed in Princeton in 1964 and in Moscow in 1971 verified EEP to 1 part in 10(12). Gravitational red shift measurements using gamma rays climbing a tower on the Harvard University campus (1965), using light emitted from the surface of the Sun (1965), and using atomic clocks flown in aircraft and rockets (1976) have verified that effect to precisions of better than 1 percent.

General Relativity

The principle of equivalence and its experimental confirmation reveal that space-time is curved by the presence of matter, but they do not indicate how much space-time curvature matter actually produces. To determine this curvature requires a specific metric theory of gravity, such as general relativity, which provides a set of equations that allow computation of the space-time curvature from a given distribution of matter. These are called field equations. Einstein's aim was to find the simplest field equations that could be constructed in terms of the space-time curvature and that would have the matter distribution as source. The result was a set of 10 equations. This is not, however, the only possible metric theory. In 1960, C. H. Brans and Robert Dicke developed a metric theory that proposed, in addition to field equations for curvature, equations for an additional gravitational field whose role was to mediate and augment the way in which matter generated curvature. Between 1960 and 1976 it became a serious competitor to general relativity. Many other metric theories have also been invented since 1916.

An important issue, therefore, is whether general relativity is indeed the correct theory of gravity. The only way to answer this question is by means of experiment. In the past scientists customarily spoke of the three classical tests proposed by Einstein: gravitational red shift, light deflection, and the perihelion shift of Mercury. The red shift, however, is a test of the equivalence principle, not of general relativity itself, and two new important tests have been discovered since Einstein's time: the time-delay by I. I. Shapiro in 1964, and the Nordtvedt effect by K. Nordtvedt, Jr., in 1968.

The confirmation of the deflection of starlight by the Sun by the solar eclipse expedition of 1919 was one of the triumphant moments for general relativity and brought Einstein world-wide fame. According to the theory, a ray of light propagating through the curved space-time near the Sun should be deflected in direction by 1.75 seconds of arc if it grazes the solar surface. Unfortunately, measurements of the deflection of optical starlight are difficult (in part because of the need for a solar eclipse to obscure the light of the Sun), and repeated measurements between 1919 and 1973 yielded inaccurate results. This method has been supplanted by measurements of the deflection of radio waves from distant quasars using radio-telescope interferometers, which can operate in broad daylight. Between 1969 and 1975, 12 such measurements ultimately yielded agreement, to 1 percent, with the predicted deflection of general relativity.

The time-delay effect is a small delay in the return of a light signal sent through the curved space-time near the Sun to a planet or spacecraft on the far side of the Sun and back to Earth. For a ray that grazes the solar surface, the delay amounts to 200 millionths of a second. Since 1964, a systematic program of radar ranging to the planets Mercury and Venus, to the spacecraft Mariners 6, 7, and 9, and to the Viking orbiters and landers on Mars has been able to confirm this prediction to better than half of 1 percent.

The Nordtvedt effect is one that does not occur in general relativity but is predicted by many alternative metric theories of gravity, including the Brans-Dicke theory. It is a possible violation of the equality of acceleration of massive bodies that are bound by gravitation, such as planets or stars. The existence of such an effect would not violate the weak equivalence principle that was used as a foundation for curved space-time, as that principle applies only to modest-sized objects whose internal gravitational binding is negligible. One of the remarkable properties of general relativity is that it satisfies EEP for all types of bodies. If the Nordtvedt effect were to occur, then the Earth and Moon would be attracted by the Sun with slightly different accelerations, resulting in a small perturbation in the lunar orbit that could be detected by lunar laser ranging, a technique of measuring the distance to the Moon using laser pulses reflected from arrays of mirrors deposited there by Apollo astronauts. In data taken between 1969 and 1976, no such perturbation was detected, down to a precision of 30 cm (1 ft), in complete agreement with the zero prediction of general relativity and in disagreement with the prediction of the Brans-Dicke theory.

A number of secondary tests of more subtle gravitational effects have also been performed during the last decade. General relativity has passed every one, while many of its competitors have failed. Tests of gravitational radiation and inertial frame-dragging are now being devised. One experiment would involve placing spinning objects in Earth orbit and measuring expected relativistic effects.

Cosmology and Time

One of the first astronomical applications of general relativity was in the area of cosmology. The theory predicts that the universe could be expanding from an initially condensed state, a process known as the big bang. For a number of years the big bang theory was contested by an alternative known as the steady state theory, based on the concept of the continuous creation of matter throughout the universe. Later knowledge gained about the universe, however, has strongly supported the big bang theory as against its competitors. Such findings either were predicted by or did not conflict with relativity theory, thus also further supporting the theory. Perhaps the most critical piece of evidence was the discovery, in 1965, of what is called background radiation. This "sea" of electromagnetic radiation fills the universe at a temperature of about 2.7 K (2.7 degrees C above absolute zero). Background radiation had been proposed by general relativity as the remaining trace of an early, hot phase of the universe following the big bang. The observed cosmic abundance of helium (20 to 30 percent by weight) is also a required result of the big-bang conditions predicted by relativity theory.

In addition, general relativity has suggested various kinds of celestial phenomena that could exist, including neutron stars, black holes, gravitational lenses, and gravitational waves. According to relativistic theory, neutron stars would be small but extremely dense stellar bodies. Aneutron star with a mass equal to that of the Sun, for example, would have a radius of only 10 km (6 mi). Stars of this nature have been so compressed by gravitational forces that their density is comparable to densities within the nuclei of atoms, and they are composed primarily of neutrons. Such stars are thought to occur as a by-product of violent celestial events such as supernovae and other gravitational implosions of stars. Since neutron stars were first proposed in the 1930s, numerous celestial objects that exhibit characteristics of this sort have been identified. In 1967 the first of many objects now called pulsars was also detected. These stars, which emit rapid regular pulses of radiation, are now taken to be rapidly spinning neutron stars, with the pulse period representing the period of rotation.

Black holes are among the most exotic of the predictions of general relativity, although the concept itself dates from long before the 20th century. These theorized objects are celestial bodies with so strong a gravitational field that no particles or radiation can escape from them, not even light—hence the name. Black holes most likely would be produced by the implosions of extremely massive stars, and they could continue to grow as other material entered their field of attraction. Some theorists have speculated that supermassive black holes may exist at the centers of some clusters of stars and of some galaxies, including our own. While the existence of such black holes has not been proven beyond all doubt, evidence for their presence at a number of known sites is very strong.

In theory, even a relatively small mass could become a black hole. The mass would have to be compressed to higher and higher densities until it diminished to a certain critical radius, the so-called "event horizon," named the Schwarzschild radius because it was first calculated in 1916 by German astronomer Karl Schwarzschild. (His calculations apply to a nonrotating object. The figures for a rotating object were developed in 1963 by New Zealand mathematician Roy Kerr.) For an object having the mass of the Sun the event horizon would be approximately 3 km (2 mi). Scientists such as the English theoretical physicist Stephen Hawking have speculated that tiny black holes may indeed exist.

The concept of gravitational lenses is based on the already discussed and proven relativistic prediction that when light from a celestial object passes near a massive body such as a star, its path is deflected. The amount of deflection depends on the massiveness of the intervening body. From this came the notion that very massive celestial objects such as galaxies could act as the equivalent of crude optical lenses for light coming from still more distant objects beyond them. An actual gravitational lens was first identified in 1979.

One phenomenon predicted by general relativity has not yet been substantially verified, however: the existence of gravitational waves. Gravitational waves would be produced by changes in gravitational fields. They would travel at the speed of light, transport energy, and induce relative motion between pairs of particles in their path (or produce strains in more massive objects). Astrophysicists think that gravitational waves should be emitted by dynamic sources such as supernovae, massive binary (or multiple-star) systems, and black holes or collisions between black holes. Various attempts, unsuccessful thus far, have been made to observe such waves.

A more fundamental matter confronting general relativity is that of the attempt being made by physicists to unite gravitation with quantum mechanics, the other paradigm of modern physics. This search for some unified field theory is the major task of workers in quantum cosmology.

The Clock Paradox

The clock paradox is the best-known example of a class of paradoxes that have been devised in order to test the logical consistency of the special theory of relativity or, in some cases, in attempts to discredit the theory. To formulate such paradoxes, one imagines a "thought experiment" for which the theory apparently makes one prediction when the experiment is analyzed one way, but makes a contradictory prediction when it is analyzed a different way. It is generally agreed, however, that all such paradoxes can be resolved quite simply by a correct application of the principles of relativity.

A simple version of the clock paradox is stated as follows: two identical twins, Charlie and Kip, part company, Charlie remaining at home and Kip piloting a rocket ship at constant velocity to a distant location. Upon reaching his destination, Kip reverses direction and returns home at constant velocity. From Charlie's point of view, Kip's clocks have run more slowly than his because of the relativistic time dilation resulting from Kip's velocity; therefore Kip should return younger than his former twin. This is a surprising prediction and, although true, it is not in itself paradoxical. The apparent paradox occurs when the situation is analyzed from Kip's point of view. From Kip's viewpoint, it is Charlie who moves with constant velocity away and then back. Therefore his clocks should run more slowly than Kip's, and it is Charlie who should be younger. Both predictions cannot be correct, hence the apparent paradox.

The simplest way to resolve the paradox is to realize that the two situations are not really symmetrical. Charlie remains in an inertial frame (or nonaccelerating environment, in which Newton's laws hold) throughout the experiment, whereas Kip must undergo an acceleration in order to halt his rocket ship and reverse its direction. During this acceleration, he notes that Charlie's clocks speed up and in fact overtake his clocks. This he understands by carefully carrying out special relativistic calculations using a sequence of inertial frames that at each instant during his acceleration are momentarily at rest with respect to him. Because these frames have different velocities, he must take into account the systematic difference in the way that clocks are synchronized from one frame to the next.

An alternative way to understand the speedup of Charlie's clocks is to realize that because of the principle of equivalence, Kip cannot distinguish his accelerated frame from one at rest in a uniform gravitational field, in which there is an apparent "gravitational blueshift" increasing the rates of Charlie's clocks. The final conclusion: both observers agree that Kip must return younger than Charlie.

An experiment performed in 1966 in an accelerator at CERN in Geneva, Switzerland, verified this result, in which the travelers were unstable elementary particles called mu mesons confined by means of magnetic fields to move in circular paths with velocities 99.6 percent the speed of light. The returning muons were found to be younger—that is, to have decayed more slowly, than muons at rest in the laboratory. Thus, both experiments and a correct application of theory verify that there is no clock paradox.

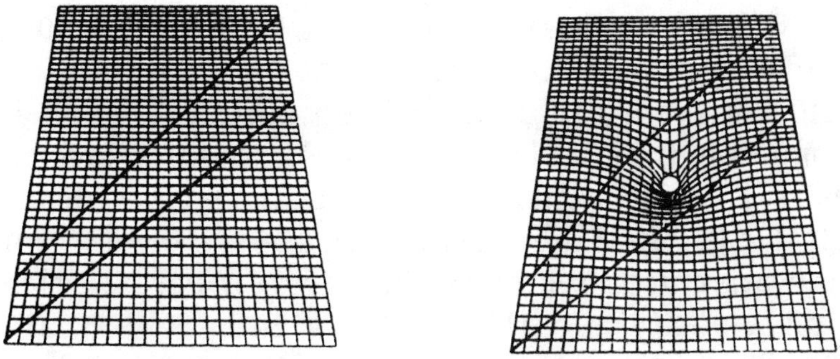

The Figure illustrates the curvature of space-time with a two dimensional plane representing three dimensional space. The lines are the path light takes as it travels through space. A massive body warps space, causing the path of light to bend. Energy also causes space to curve, as Einstein's theory of general relativity describes.
From *Tapping the Zero Point Energy* by Moray B. King.

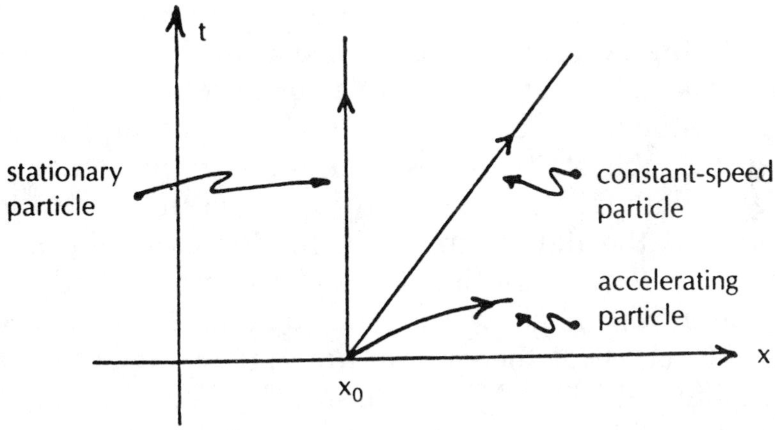

World lines of three particles.

From *Time Machines* by Paul Nahin.

Space-Time Continuum

As mentioned above, the space-time continuum is a concept associated with the theory of relativity. It replaced the Newtonian concept of space and a separate, absolute time. In Newtonian mechanics, any event can be associated with a location in space and with a moment in time t. The coordinates chosen are arbitrary, but two quantities are independent of the choice: the spatial distance between two events, Delta I, and the difference in time between them, Delta t.

With the advent of relativity, however, it became clear that time depends on velocity and that Delta I and Delta t are no longer separately invariant. Delta I undergoes the Fitzgerald-Lorentz contraction and Delta t undergoes time dilation. A new quantity, Delta s, is invariant instead. This quantity, known as the "line element" or "invariant interval," is related to the other quantities through a quadratic expression involving the speed of light. The quantity Delta s is now the invariant measure of intervals between events, and the term metric (from the Greek word for "measure") often refers to the quadratic expression for Delta s(2). In general relativity the space-time metric is more complicated and corresponds to curved space-time.

The Fitzgerald-Lorentz contraction is an effect first postulated in 1892 by George F. Fitzgerald and amplified in 1895 by Hendrik A. Lorentz. It was proposed in an attempt to explain the null result of the Michelson-Morley experiment, performed in 1887, which measured the time taken for a light beam to travel a distance d and back when the direction of motion of the beam was parallel to the supposed direction of motion of the laboratory, or observer, through the "ether." The time was compared with that for the same trip when the laboratory moved perpendicular to the light's direction. In this case, the laboratory's motion was the Earth's motion through space. The ether was the medium through which light was presumed to travel with speed c.

According to classical theory, the time taken for both round trips should be different. The Michelson-Morley experiment, however, demonstrated no difference in travel times. Fitzgerald suggested that if the length of the parallel arm d were contracted to d X the square root of (1 - v(2)/c (2)) (where v is the speed of the laboratory) while that of the perpendicular arm remained unchanged, the prediction would agree with the experiment.

Lorentz later proposed a model for matter that incorporated this effect. He stated that the atoms and molecules that compose matter would, under the effect of motion, be compressed along the direction of motion. It would be impossible to measure this deformation; a ruler placed alongside a speeding object would be similarly shortened.

Albert Einstein showed, in his theory of relativity, that the contraction was a fundamental consequence of the assumption that the speed of light is the same in all reference frames. The effect is only significant at relativistic speeds, or speeds that are a sizable fraction of the speed of light.

The World Line

A world line is a path through space-time. It is a mathematical concept used in physics to describe the motion of particles or other objects. A world line gives a more informative description than a path through space, because each point along the world line indicates both the time and the position of the object. To understand the distinction, consider a curve drawn on a map to indicate the path of an aircraft over the surface of the Earth. If the altitude is marked along the curve, the path is described in three-dimensional space. If the time that the aircraft passes each point is also indicated, the description is of the world line, or path, in the four-dimensional space-time continuum. Two world lines intersect only if they cross the same position at the same time.

Physicists use world lines to describe how particles move, collide, and recoil from each other by their fundamental interactions. Metric theories of gravitation, for example, such as Einstein's general relativity, predict that freely falling particles move along geodesics, the paths closest to straight lines in curved space-times. Real particles with nonzero rest masses travel slower than light and move on timelike geodesics, which give the longest time measured between two events. Other world lines between the same two events experience less elapsed time by the clock paradox.

Time Reversal Invariance

A motion in which events appear in reverse order is said to be time reversed with respect to the original motion. If some motion is possible according to known physical laws, then the time-reversed motion is almost always possible. This possibility is true for motions caused by gravitational and electromagnetic forces. It is therefore said that these laws exhibit time reversal invariance, or have time reversal as a basic symmetry. For example, it would be unusual to observe a real process in which an egg broken on the floor collects itself and flies up whole into a person's hand, as in a movie run backward. Nevertheless, according to known physical laws such a process is not impossible, although it is too unlikely to expect it actually to happen. Movies of simpler processes, however, such as a spinning top or a vibrating bell, appear nearly normal when run backward.

The basic natural laws governing most elementary processes are invariant under time reversal. Most natural processes themselves, however, do not appear to be symmetric under time reversal. This result is because of the complexity of macroscopic systems containing many particles. For example, consider a vessel divided into two parts by an airtight partition. One part contains air and the other is evacuated. If the partition is removed, air will expand until both parts of the vessel are filled. To obtain the time-reversed motion in which all the air filling the vessel streams to one side, the velocities of all the molecules would have to be reversed—an impossible task. The complexity of this system results in an extremely small likelihood that the time-reversed motion could occur. Complex systems generally develop in time toward a more probable state. This increase in likelihood is related to an increase of entropy of the system.

On the subatomic level there are additional forces to consider with respect to time reversal invariance—the strong nuclear force responsible for binding atomic nuclei and the weak force responsible for the radioactive decay of such particles as the neutron. The neutron decays into a proton, an electron, and an antineutrino. The motion that is time reversed with respect to the neutron decay is represented by a formula in which an antineutrino, an electron, and a proton come together to form a neutron. In order for the time-reversed reaction to correspond to the original decay, the energies of the incoming particles must be the same as those of the previous outgoing particles, and the velocities and spins of the particles must be reversed. This process is possible, although unlikely.

Experiments with elementary particles test time reversal invariance by observing processes in which elementary particles are produced, interact, and decay. Time reversal invariance appears to be a valid symmetry for all processes except in the weak decay of K mesons (kaons).

Time Travel and the Quantum Blob

In the fall of 1998 it was reported in the world-wide press that a special experiment had taken place at Sussex University in the England that literally "froze" time.

Said newspapers in London at the time, "For a few seconds last month the coldest spot in the universe was in Brighton. While the temperature outside hovered around 20C on September 22, inside Malcolm Boshier's optics lab at Sussex University it was a few hundred billionths of a degree above the coldest temperature possible, -273C.

"When Boshier and his team dropped 100,000 atoms down to nanokelvin temperatures, they were not merely trying to get into the *Guinness Book of Records*. This is a step on the road to making the world's most sensitive measuring instruments, so sensitive they might help to unlock the secrets of gravity.

"The big freeze created a Bose-Einstein Condensate (BEC) which occurs when atoms lose almost all their energy. In this state an extraordinary thing happens: each atom loses its individuality. They spread out into each other, forming a blob that behaves like one enormous 'superatom.' "It's an object that obeys the laws of quantum mechanics, yet it's large enough that you can take a picture of it," says Boshier.

"Albert Einstein and Satyendra Bose predicted such a thing would happen at ultra-low temperatures more than 70 years ago, but it wasn't until 1995 that a Colorado research group first made a BEC. "This represents the tightest control you can have over atoms," says Boshier. "We should be able to build devices that will be extremely sensitive to anything that affects an atom's energy levels, and that includes gravity."

"An understanding of just how gravity works still eludes researchers, and an instrument to measure its subtle effects will be vital in unearthing its secrets. The next stage is about to begin. Sussex's Centre for Optical and Atomic Physics has already produced 'magnetic mirrors,' which bounce atoms around in the same way that mirrors bounce laser light. Boshier will soon be dropping the BEC onto a magnetic mirror to check that it bounces like normal atoms.

"If it does, the team will try to build an 'atom interferometer.' Lasers provide highly accurate interferometers because the photons of the tiny packets of light in their beams are locked in step. This provides a way to measure tiny dimensions. Split a laser beam in two, and the relative length of the two different beam paths can be [measured. Those] in one half are slightly out of step with the others. This has led to highly sensitive instruments such as laser gyroscopes.

"In a BEC, the atoms, like the photons in a laser beam, are locked together, which will provide an even more accurate interferometer. Creating a BEC involves two stages. First put rubidium vapour into an evacuated flask. Six lasers are arranged around the flask with their beams intersecting at the centre. To a normal, moving atom this is like wading through what the researchers call 'optical molasses.' The atom loses energy, slows down, and cools to around one thousandth of a degree above absolute zero, or -273C; cold, but not cold enough.

"The lasers are constantly adding energy to the atoms, so they are turned off and a specially-shaped magnetic field is turned on. This holds the atoms as if it were a deep-sided bowl. The atoms with the most energy steal some from the others and escape from the top of this magnetic trap: thus the remaining atoms get colder.

"Eventually an atom reaches the point at which it can move into its lowest quantum state: the ground state. The nature of these atoms is such that others quickly follow. "When it is cold enough so that there is some chance of having one atom in the ground state, many of the others scatter into that same ground state," says Boshier.

"The result is a quantum blob a few microns in diameter. It may not sound like much, but it contains the future, Boshier believes."

Boshier is then quoted as saying, "This field is really exploding: five or six hundred papers—most of them theoretical—have been written on it in the last three years. This is going to do for atoms what lasers did for light."

Indeed, some of the secrets of time travel may have been unlocked by Boshier and his team. The technique of splitting a laser beam in two and the use of laser gyroscopes may ultimately lead us to the practical time travel device that many are looking for, as we shall see in later chapters.

Meanwhile, let us look back in time, a hundred years or so, at the beginning of our own era's interest in time travel. For it seems that those who are interested in time travel are likely to meet time travellers!

The material in this chapter is largely from the *Grolier Multimedia Encyclopedia*.[39] Bibliographies given are as follows:

Bibliography: Thorne, Kip, *Black Holes and Time Warps* (1994); Will, Clifford, *Was Einstein Right?* (1987).

Sachs, Robert G., *The Physics of Time Reversal* (1987).

Friedman, M., *Foundations of Space-Time Theories* (1983); Laurent, B., *Introduction to Spacetime* (1995); Morris, R., Time's Arrows (1985); Ray, C., Time, *Space and Philosophy* (1991); Wald, R., *Space, Time, and Gravity, 2d ed.* (1992); Zeh, H., *The Physical Basis of the Direction of Time, 2d ed.* (1992).

Taylor, E. F., and Wheeler, J. A., *Spacetime Physics* (1966).

Childress, D. ed., *Anti-Gravity & the Unified Field* (1990).

A drawing of the 1896 California airship that was widely circulated in newspapers at the time. Note the powerful searchlight that the airship was described as having.

Descriptions of the "airships" were within the general framework of "cigar-shaped" and "metallic" but varied as to the kinds of appendages seemingly attached to the body: wings, propellers, and fins were all described by different witnesses. Both daytime and nighttime sightings were reported, and in the case of the latter, brilliant lights were seen.

Reports of sightings continued through May 1897 and came from almost every state of the Union east of the Rockies, with a concentration in the Midwest. (*The Encyclopedia of UFOs*, Ronald D. Story, ed., pp.8–11)

The Call

SAN FRANCISCO, SUNDAY MORNING, NOVEMBER 22, 1896—TWENTY-EIGHT PAGES.

SAW THE MYSTIC FLYING LIGHT

Oaklanders Who Believe an Airship Hovered Over Them.

Say That They Saw a Dark Body Above the Gleam.

It Was Headed for San Francisco and Seemed About to Come to Earth.

OAKLAND OFFICE SAN FRANCISCO CALL, 908 Broadway, Nov. 21.

A great wave of UFO sightings occurred in California in the late 1800's. This story and illustration appeared in a San Francisco newspaper on November 22, 1896.

Another wave of "airships" was reported from New Zealand during 1909. For six weeks, from late July to early September, hundreds of people observed cigar-shaped airships over North Island and South Island. Sightings occurred in daytime and nighttime.

The early 20th century was a time of experimental aircraft. The dirigible balloons being built by Graf Ferdinand von Zeppelin in Germany were still in an early stage of development, and so were the balloons being tested in France. These craft could not possibly have found their way to New Zealand.

3.
Time Bandits from the 1800s

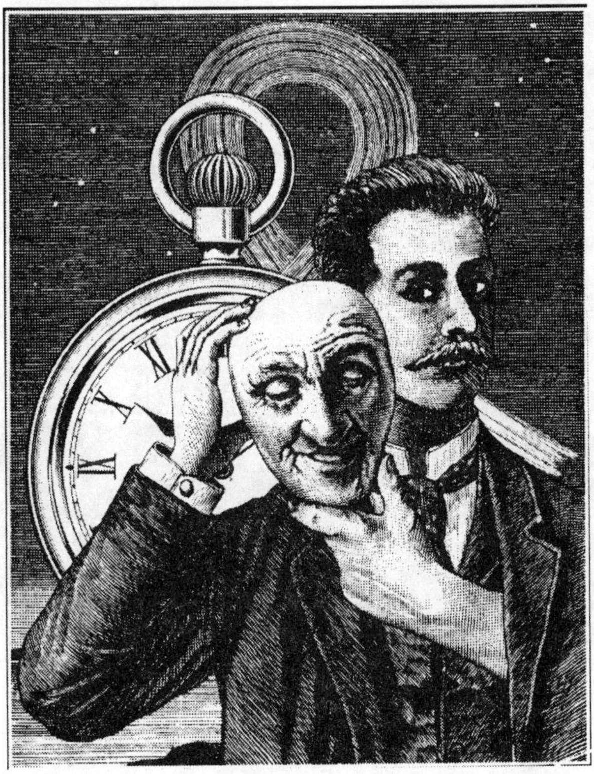

An illustration of the 'Clipper of the Clouds' from the 1886 Jules Verne book *Robur le Conquerant*.

3. Time Bandits from the 1800s

The miracle is not to fly in the air,
or to walk on the water,
but to walk on the earth.
—*Chinese proverb*

Man is most nearly himself
when he achieves the seriousness of a child at play.
—*Heraclitus,* Greek philosopher (500 B.C.)

Time travel as a modern product, especially if allowed to the consumer, would probably soon come under some form of abuse. Time Bandits, travelling back into time of personal profit, would eventually necessitate the need of some sort of time police. This subject has been explored in various movies including the George Harrison film *Time Bandits* and the Jean Claude van Damme film *Time Cop.* Many, many films and television series have featured time travel as their basic plots, including *The Time Tunnel* (1968), *Sliders* (1986), the British "Dr. Who" series, the *Back to the Future* films and many other productions. Indeed, the theme of time travel is definitely popular with Hollywood.

Some may say that it all started with George Pal's film version of H.G. Wells' 1895 novel, *The Time Machine.* The 1960 MGM film starred Rod Taylor and featured a wonderfully Victorian gadget that eventually took the hero to A.D. 802,701. Others theorize that the invention of the time machine (at least for consumer purposes) in 2112 A.D. really sparked the whole time travel craze, with a certain altering of history as a result, including the retroactive writing of books and the making of movies and television shows. How boring the future would have been without all the time travellers. They decided that a certain acceleration of history and invention was necessary.

Where Have All the Time Travellers Gone?

Skeptics of the time travel phenomenon point out that if time travel were a reality, we would hear much more about time travelers, on talk shows and such. Arthur C. Clarke wrote in 1985, "The most convincing argument against time travel is the remarkable scarcity of time travellers. However unpleasant our age may appear to the future, surely one would expect scholars and students to visit us, if such a thing were possible at all. Though they might try to disguise themselves, accidents would be bound to happen—just as they would if we went back to Imperial Rome with cameras and tape-recorders concealed under our nylon togas. Time travelling could never be kept secret for very long."

Physicist Paul Nahin in his scholarly work *Time Machines* (published by the American Institute of Physics) quotes from physicist McDevitt: "If it [time travel] could be done, someone would eventually learn how. If that happens, history would be littered with tourists. They'd be everywhere. They'd be on the Santa Maria, they'd be at Appomattox with Polaroids, they'd be waiting outside the tomb, for God's sake, on Easter morning."[6]

Nahin, Clarke, McDevitt and others want to know, where are all the time travellers? Says Nahin, "From the moment after the first time machine is constructed, through the rest of civilization, there would be numerous historians (to say nothing of weekend sight seekers) who would want to visit every important historical event in recorded history. They might each come from a different future, but all would arrive at destinations crowded with temporal colleagues—crowds for which there is no historical evidence!"[6]

Nahin quotes another physicist named Fulmer who commented, "Actually I know of only one argument against the possibility of time travel that seems to carry any weight at all. This is the fact that it does not appear ever to have happened. That is, it might be argued that there will be no time trips from 1985 to 1975, since we were here in 1975 and saw no time travelers. But this argument is far from conclusive."[6]

Indeed, just because we haven't seen a time traveller on the *Tonight Show* as a guest doesn't mean that they are not among us right now. It would be dangerous to travel to many ancient periods, and many historical events that one might want to visit, such as battles, assassinations, and such, would be especially hazardous. Nonetheless, time travellers may have risked death ("Sorry, but your son died in an accident 2,000 years ago while time travelling..."), and historical reports do show a high incidence of "unusual" people and "odd" incidents that might be best explained in terms of time travel.

Volumes of books could be filled with incidents from history that may have something to do with time travel. I will mention but a few here. Nearly all of the "ancient astronaut" evidence that can be found in the hundreds of books on the subject, can be alternatively explained in the time travel hypothesis, and have been. British writer Robin Collins covered this ground in his *Ancient Astronauts: A Time Reversal*[3] in 1976 at the height of the "Gods from Outer Space" craze of the early 1970s.

There are various odd accounts of visitations from very human-looking "angels" who warn people (such as Lot, to leave Sodom and Gomorrah) before a sudden cataclysm. Knowledge of "future" events is evidence of a kind for time travel. Similarly, many historical events often have some element of premonition and "warning" accompanying them. The participants cannot help but complete their parts in the script, but many observers "know" what is going to happen. "We warned you," is a commonly heard phrase. Time travellers must use it a lot.

There are certain historical personages, the famous Count de Saint Germain for example, who seem to have been time travelling geniuses who came and went as they chose. In some cases, perhaps they even faked their deaths at the appropriate time, much like the hero of *The Highlander* television series; or sometimes just faded away into the foggy mists of history. Some characters such as Enoch or Saint Germain were never known to have actually died. Neither was the Chinese philosopher Lao Tzu who wrote *The Tao Te Ching,* probably ancient China's most famous book.

Readers of Charles Fort's books or the British publication, *Fortean Times,* will already be familiar with reports of strange dimension-blurring occurrences, from mysterious visitors who seemingly know everything (including the future), to falls of various fishes or frogs from the clear blue sky.

As the media becomes faster and more omnipresent, stories of time travellers may begin to appear—and may be quickly discounted, but at least they would be reported. As to ancient history, very little description has really come down to us. Starting in the early 1800s, countries with advanced newspaper industries such as England, Germany, France and America, started seeing reports of high-tech aerial machines and mysterious men with strange devices (including flying machines akin to UFOs), that interacted briefly with the locals and then disappeared. Unlike similar events in the past, these unusual events actually made it to newspapers and were archived for research in the future!

The Mysterious Airships of the Late 1800s: Time Travellers?

Mysterious airships were apparently crossing the Atlantic Ocean over 100 years ago. William Corliss of the Sourcebook Project mentions in his *Unexplained Phenomena* calendar for 1999 that the captain and crew of the *Lady of the Lake,* a British steamer in the the Atlantic Ocean off the coast of West Africa, south of Cape Verde, observed a strange sight on March 22, 1870. It was a gray object divided into four connected sections. Behind it trailed a long "hook" connected to the center of the UFO. Positioned below the clouds, it flew against the wind and was visible for half an hour.

In 1873 at Bonham, Texas, workers in a cotton field suddenly saw a shiny, silver object in the sky that came streaking down at them. Terrified, they ran away, while the "great silvery serpent" as some people described it, swung around and dived at them again. A team of horses ran away and the driver was thrown beneath the wheels of the wagon and killed. A few hours later that same day in Fort Riley, Kansas, a similar "airship" swooped down out of the skies at a cavalry parade and terrorized the horses to such an extent that the cavalry drill ended in a tumult.

Airships, often with powerful searchlights at their front, plied the skies of North America and other continents during the 1880s and 1890s and finally culminated in a huge wave of sightings in 1897.

These sightings and contacts started in November, 1896 in San Francisco, California when hundreds of residents saw a large, elongated, dark object that used brilliant searchlights and moved against the wind, traveling northwest across Oakland. A few hours later reports came from other northern California cities such as Santa Rosa, Chico, Sacramento and Red Bluff all describing what appears to be the same airship, a cigar shaped craft. It is quite possible that this craft was heading for Mount Shasta in northern California.

The airship moved very slowly and majestically, flying low at times, and at night, shining its powerful searchlight on the ground. It is worth noting here, as Jacques Vallee did in his book *Dimensions*,[25] that the airship could do exactly as it cared to, because unlike today, it ran no risk of being pursued. There were no jet squadrons to be scrambled after the aerial intruder, nor anti-aircraft guns or surface to air missiles to shoot down this trespassing craft in the sky.

However, the airship, clearly not a typical balloon or gas-filled airship of the time, did at times move erratically; sometimes it would depart "as a shot out of a gun," change course abruptly, change altitude at great speed, circle and land and, as previously mentioned, use powerful searchlights to sweep the countryside.[25]

These mysterious airships were seen across the United States, from California to Nebraska, Texas, Colorado, Kansas, Iowa, Missouri, Wisconsin and Minnesota, including many heavily populated urban areas such as Omaha and Milwaukee. On April 10, 1897, thousands of people in Chicago reported seeing a cigar shaped airship.

Jerome Clark reports in *The UFO Encyclopedia* that on February 1, 1897 the *Omaha Daily Beef* ran a story of a "large, glaring light" which hovered, ascended, descended, and moved at a "most remarkable speed," over Hastings, Nebraska.[24]

He also states that the *St. Louis Dispatch* reported on April 16, 1897 that a St. Louis man claimed to encounter an airship and "Martian couple" in the hills outside of Springfield, Missouri. He said that they did not speak to him and that they were handsome, long haired, and friendly, and they did not wear clothes. Says Clark, "Apparently they did not understand the concept of time because they were puzzled by his watch."

It is generally agreed that the many accounts of these airships could not be attributed to known airships or technology of the time. The first powered flight was Giffard's steam airship built in 1852, while the Tissandier brothers built the first electric airship in 1883. Renard and Kreb's electric airship, the *La France*, was first flown at Chalais-Meudon in 1884. The Schwartz aluminum rigid airship was first flown at Tempelhofer Field, Germany, in 1897 and the first "successful" airship, the *Lebaudy* was test flown in Paris in 1903.[19]

A great deal has been made of the airship flap of 1897 in UFO circles, typically seeking to prove that the airships were extraterrestrial vehicles. Yet, as Jacques Vallee points out in *Dimensions,*[25] the evidence does not point toward extraterrestrial occupants because those airship operators who engaged in conversation with witness "were indistinguishable from the average American population of the time."

Many occupants did indeed converse with locals and occasionally offered them rides, though few seemed to have ever taken up the offers. In a curious incident at Hot Springs, Arkansas (a site of natural crystals) on the night of May 6, 1897, Constable Sumpter and Deputy Sheriff McLemore witnessed an airship land on a rainy night. Drawing their Winchesters they demanded an occupant to identify himself and the airship. A man with a long dark beard came forth with a lantern in his hand (possibly electric?) and "on being informed who we were proceeded to tell us that he and the others—a young man and a woman—were traveling through the country in an airship. We could plainly distinguish the outlines of the vessel, which was cigar-shaped and about sixty feet long, and looking just like the the cuts that have appeared in the papers recently. It was dark and raining and the young man was filling a big sack with water about thirty yards away, and the woman was particular to keep back in the dark. She was holding an umbrella over her head. The man with the whiskers invited us to take a ride, saying that he could take us where it was not raining. We told him we believed we preferred to get wet.

"Asking the man why the brilliant light was turned on and off so much, he replied that the light was so powerful that it consumed a great deal of his motive power... Being in a hurry we left and upon our return, about forty minutes later, nothing was to be seen. We did not hear or see the airship when it departed."[25]

In another fascinating and revealing report, this one from the *Houston Post* of April 22, 1897, a Mr. John M. Barclay living near Houston witnessed an airship on the ground on the night of April 21 at about 11:00 P.M. "It was a peculiar shaped body, with an oblong shape, with wings and side attachments of various sizes and shapes. There were bright lights, which appeared much brighter than electric lights."

When he first saw it, it seemed perfectly stationary about five yards from the ground. It circled a few times and gradually descended to the ground in a pasture adjacent to his house. He took his Winchester and went down to investigate. As soon as the ship, or whatever it might be, alighted, the lights went out.

The night was bright enough for a man to be distinguished several yards away, and when within about thirty yards of the ship he was met by an ordinary mortal, who requested him to lay his gun aside as no harm was intended. Whereupon the following conversation ensued. Mr. Barclay inquired: "Who are you and what do you want?"

The reply by this strange individual was, "Never mind about my name, call it

Smith. I want some lubricating oil and a couple of cold chisels if you can get them, and some bluestone. I suppose the saw mill hard by has the two former articles and the telegraph operator has the bluestone. Here is a ten-dollar bill: take it and get us these articles and keep the change for your trouble."

Mr. Barclay said, "What have you got down there? Let me go and see it."

He who wanted to be called Smith said: "No, we cannot permit you to approach any nearer, but do as we request you and your kindness will be appreciated, and we will call you some future day and reciprocate your kindness by taking you on a trip."

Said the newspaper reports, "Mr. Barclay went and procured the oil and cold chisels, but could not get the bluestone. They had no change and Mr. Barclay tendered him the the ten-dollar bill, but same was refused. The man shook hands with him and thanked him cordially and asked that he not follow him to the vessel. As he left Mr. Barclay called him and asked him where he was from and where he was going. He replied, "From anywhere, but we will be in Greece day after tomorrow." He got on board, when there was again the whirling noise, and the thing was gone, as Mr. Barclay expresses it, like a shot out of a gun. Mr. Barclay is perfectly reliable."[20, 25]

While such incidents are baffling to most people, in the context of time travellers they do not seem so extraordinary. Jacques Vallee thinks the statement from the stranger that he is "From anywhere, but we will be in Greece day after tomorrow," is absurd. Yet what is absurd about a vague answer as to one's origins and that he plans to be in Greece in two days? In 1897, by the airship technology of the time, this was impossible; and what extraterrestrial, asks Vallee, would state that they would be in Greece in two days? For a human time traveller on his way to Greece, however, the answer seems quite sensible.

The airship wave of 1896/1897 will never be fully solved. Does it involve time travellers? Records from the future tend to confirm this thesis. Of the 100 or so reported sightings across the country, some were obvious hoaxes and fabrications based on the many newspaper articles appearing at the time. Yet, with those genuine sightings, considerable doubt remains as to the nature of these craft. Says Wallace Chariton at the end of his book *The Great Texas Airship Mystery*[20]: "Many 1897 witnesses said they heard a peculiar whirring or whizzing sound that could not be identified. There were several reports that the flying machine hovered in one spot for some time then quickly disappeared traveling at a high rate of speed. There was always at least one light that was frequently described as being the brightest light the witnesses had ever seen and was often said to be considerably more powerful than any incandescent light, which was the only kind they had in 1897. Some witnesses said they saw a bright, fluorescent glow about the ship and many others claimed there were multicolored lights along the sides. If you do any research into reported modern UFO sightings you will find that similar statements occur frequently."

The 1895 Time Machine

Just prior to the airship flap of 1897, a remarkable novel was published, *The Time Machine.* This 1895 book by H. G. Wells was the first of the inventive scientific fantasies on which he built his early reputation. In the story, a wealthy inventor invents a time machine and travels into the future, stepping out of his machine into what is left of his London basement laboratory.

Like Wells' later sociophilosophical works, *The Time Machine* is a serious story with a political message concerning future relations between the ruling and working classes in industrial society. The Time Traveller of his story travels by machine to an era in which the world is inhabited by two races, the pleasure-loving Eloi and the ape-like Morlocks who live underground and feed on the Eloi. As Wells' ironic comment on the outcome of 19th-century progress, *The Time Machine* is a gloomy and pessimistic parable of the future of humankind.

Was H. G. Wells somehow introduced to the concept of a time machine by time travellers themselves? This was also around the time of the Wilson Brothers of EMI/Thorn and their alleged connection with the Philadelphia Experiment. EMI/Thorn is British media group, based in London, that goes back to the late 1800s. Though it is not as powerful a media group as it once was, EMI/Thorn is one of the oldest continuing entertainment companies with links back to the early days of Tesla, Westinghouse, Edison and General Electric and Marconi's electronic company today known as Marconi Systems (located in the greater London area, similar to EMI/Thorn).

Back in 1898, strange things were happening. Nikola Tesla was inventing alternating current and lighting New York City with his amazing new power source. The world would be forever changed and a vast industry of electrical devices was created overnight. Tesla predicted death rays, anti-gravity, thought machines and other futuristic devices. The modern era of technology was dawning. Without Telsa's alternating current and pulsed-field technology, many important inventions, such as time travel devices would never have been invented.

Strange objects were still seen in the sky, objects seemingly from the future. Jerome Clark mentions that on March 15, 1901 a Silver City, New Mexico, newspaper, the *Enterprise,* reported that eight days earlier a local man, Dr. S. H. Milliken, had taken a photograph of a peculiar flying object. The *Enterprise* stated, "The machine as seen in the picture has the appearance of three cigar-shaped objects which seemed to be lashed together, the one hanging below the other two." Unfortunately, the paper did not reproduce this potentially historic photograph, which has now been lost, Clark says. We could then determine if we were looking at a craft from the future—or the past.[24]

Clark mentions another curious incident that took place over Indiana on March 17, 1903. On that day a "huge object like a gigantic ripe cucumber with slightly tapered end" and with two rows of windows along the side hovered over a farm in Helmer, Indiana. The witnesses were the Brosius family, who estimated the cigar-shaped object to be 100 feet long. The Brosius family said that the object zigzagged across the sky at great speed and disappeared. As Clark says, this incident occurred more than 40 years before the era of "flying saucers," and before "UFOs" became a household word. Were these time travellers from the future on a package tour of 1903 America?

Only the next year Navy personnel on the *USS Supply,* a U.S. ship off of the coast of Korea, sighted three red, egg-shaped objects "beneath the clouds." Lt. Frank Schofield reported on February 28, 1904 that as the objects were approaching the ship "... they passed above the broken clouds... The largest had the apparent area of about six suns. ...the second ...the size of two suns, and third, about the size of the sun." He further commented that the objects were "most remarkable."[24] Schofield became commander-in-chief of the Pacific Fleet in the 1930s and can be considered a reliable witness.

The *London Daily Mail* ran a story on May 20, 1909, reporting that two nights earlier, while walking along a mountain path, a Cardiff, Wales, man named C. Lethbridge, saw a "large tube-shaped construction." Inside it were "two men, in heavy fur coats. When they saw Mr. Lethbridge, they spoke excitedly to each other, in a foreign language, and sailed away."[24]

Many modern UFO researchers consider this case a probable hoax, largely because the technology involved seems impossible for the time period. These may have been time travellers and the "large tube-shaped construction" may have been their spacecraft/time travel device.

William Corliss reports that on February 9, 1913, residents near Esterhazy, Saskatchewan, reported a strange formation of lights in the heavens that resembled an "express train in the sky." The procession of bright, fiery lights, moving majestically, emerged from the northwest and continued across the sky in several successions, for more than three minutes. The lights were accompanied by rumbling, like distant thunder. Corliss says that they were also seen by professional astronomers, but their explanation of the lights being meteors leaves some doubt due to their motion and appearance (moving across the sky, instead of straight down to an impact, for instance).[26]

"Excuse Me, Can You Tell Me what Time it is?"

Closer to our own time period are other disturbingly familiar encounters with ordinary humans in unusual craft. These humans may well have been time travellers. After the first reports of flying saucers of the late 1940s, several years after the so-called Philadelphia Experiment, the year 1950 saw such books as *Behind the Flying Saucers*[29] by Frank Scully and *Flying Saucers Over Los Angeles*[30] by DeWayne Johnson.

Frank Scully in *Behind the Flying Saucers* mentions the crash of a UFO near Aztec, New Mexico, on March 25, 1948. Scully claims that the craft crashing on the rocky plateau in northern New Mexico, near Farmington, might have been "Venusian." More curiously, it was claimed that investigators recovered the bodies of 16 beings, apparently humans, who were clad in the "style of 1890." Scully also speculated that the craft "probably flew on magnetic lines of force." He did not speculate on whether these were time travellers making a disastrous side-trip from 1890. The crash at Aztec remains a highly disputed event, but the Farmington area of New Mexico was a hot bed of UFO activity in the late '40s and early '50s, as recorded by Johnson and Thomas in their book *Flying Saucers Over Los Angeles.*[30]

For instance, on August 20, 1949, the famous astronomy professor Clyde Tombaugh witnessed a large, cigar-shaped craft, with a row of lit windows along the side, from a home just outside of Las Cruces, New Mexico. Professor Tombaugh is the credited discoverer of the planet Pluto in 1930.

Similarly, at 8:30 in the evening on January 20, 1951, personnel at the Sioux City, Iowa, control tower warned a Mid-Continent Airlines DC-3 about to take off to watch out for an approaching unidentified aircraft. Once in the air, the DC-3 crew and passengers saw a light closing in on the airliner at great speed. As the light got closer the crew could see that the craft was a well-lit, cigar-shaped structure. Rows of windows were reported along the sides. The object paced the DC-3 for a short time and then shot off out of sight at great speed.[24]

A few years later, on April 7, 1956, South African contactee Elizabeth Klarer met the love of her life, a handsome man named Akon. Akon was a gorgeous human hunk who was from the planet Meton. Elizabeth claimed in her book *Beyond the Light Barrier* that when she became pregnant, she went with Akon to the planet Meton for four months. Meton was a lot like Earth, only in a different time, presumably the future. Elizabeth maintained until her death in 1994 that Akon was was her only love and that he and her son would visit her periodically.[24]

"Hey, Man, Which Way to the Sixties?"

For fun-loving time travellers, the 1960s and 70s would be popular destinations. Hallucinogenic drugs were still legal, the sexual revolution was getting started, and people were beginning to wear funny clothes and look different. This would help odd-looking time travellers blend in. Being hip isn't always easy for time travellers. Strange visitations of Men in Black and flying saucer sightings became more and more common.

At approximately 11:00 A.M. on April 18, 1961, sixty-year-old chicken farmer Joe Simonton of Eagle River, Wisconsin had a highly unusual encounter. He was attracted outside by a peculiar noise similar to "knobby tires on a wet pavement." Stepping into his yard, he faced a silvery saucer-shaped object, "brighter than chrome," which appeared to be hovering close to the ground without actually touching it. The object was about twelve feet high and thirty feet in diameter.

A hatch opened about five feet from the ground and Simonton saw three men inside the machine. One was dressed in a black two-piece suit. The occupants were about five feet tall. Smooth-shaven, they appeared to "resemble Italians." They had dark hair and skin and wore outfits with turtleneck tops and knit helmets. From the description they may have been wearing overalls, and been Rastafarians—since Rastafarians often wear such "knit helmets" (as in the film *Bonzai Buckeroo and the Eighth Dimension*). Such knit helmets may hide hair, ears, headphones, or other things.

One of the men held up a jug apparently made of the same material as the saucer. His motioning to Joe Simonton seemed to indicate that he needed water. Simonton took the jug, went inside the house, and filled it. As he returned, he saw that one of the men inside the saucer was "frying food on a flameless grill of some sort." The interior of the ship was black, "the color of wrought iron." Simonton saw several instrument panels and heard a slow humming sound, similar to the hum of a generator. When he made a motion indicating he was interested in the food, one of the men, who was also dressed in black but with narrow red trim along the trousers, handed him three cookies, about three inches in diameter and perforated with small holes.

The whole affair lasted about five minutes. Finally, the man closest to the witness attached a kind of belt to a hook in his clothing and closed the hatch in such a way that Simonton could scarcely detect its outline. Then the object rose about twenty feet from the ground before taking off straight south, causing a blast of air that bent some nearby pine trees.

Along the edge of the saucer, the witness recalls, were exhaust pipes six or seven inches in diameter. The hatch was about six feet high and thirty inches wide, and, although the object has always been described as a saucer, its actual shape was that of two inverted bowls.

Simonton later reported to two Sheriff's deputies that he ate one of the cakes, and thought it "tasted like cardboard." The United States Air Force, which examined the remaining two cakes put it more scientifically: "The cake was composed of hydrogenated fat, starch, buckwheat hulls, soya bean hulls, wheat bran. Bacteria and radiation reading were normal for this material. Chemical, infrared and other destructive type tests were run on this material. The Food and Drug Laboratory of the U.S. Department of Health, Education and Welfare concluded that the material was an ordinary pancake of terrestrial origin."[25]

To well-known UFO investigator Jacques Vallee, this case is credible, yet absurd! What sort of extraterrestrials look like ordinary humans, wear overalls (Oshkosh-by-Gosh!) and hand out perfectly ordinary buckwheat pancakes to chicken farmers in Wisconsin? Interdimensional ones, he concludes. Time travelling ones, I conclude.

It is well known that time/space distortion and strange electromagnetic effects occur around many UFO encounters. On April 4, 1966, Ronald Sullivan was driving east of Bealiba, Victoria, Australia, when he was astonished to see his headlight beams bend as if they were pieces of pipe—an effect that seemed to defy the laws of physics. Sullivan stopped immediately, got out, and saw a brilliant light flashing in a field like a searchlight. He then looked up and saw that the searchlight was attached to a UFO that was just at that moment departing the scene. Later his car and headlights were in perfect working order.

Many persons have reported that their cars have come to a complete halt for no apparent reason when a UFO hovers near to them. Sullivan's bent headlights may have had to do with dimensional distortion from a time machine suddenly entering the vicinity, distorting all fields around it. Reports of craft with powerful searchlights on them go back to the 1880s.

Another curious incident occurred on December 3, 1967, when a patrolman named Herb Schirmer, of Ashland, Nebraska, had an unusual experience that changed his life forever. After writing in his logbook that night that he "saw a flying saucer at the junction of highways 6 and 63. Believe it or not..." Schirmer reported he had seen an object with a row of flickering lights take off from the highway. The patrolman decided to follow it and drove up a dirt road toward the intense light. He tried to call the police in Wahoo, Nebraska, but his radio would not work. His car died (typical of being around a strong electromagnetic field) and he stared at the object which was metallic and football shaped and surrounded by a silvery glow. It was making a "whooshing" sound, and the lights were flickering rapidly. Legs appeared under the craft, and it landed. Schirmer wanted to drive home, but he was "prevented by something in his mind."

The occupants of the craft came toward the car. He was unable to draw his revolver. A greenish gas was shot toward the car and an occupant pulled a small object from a holster, flashed a bright light at him, and he passed out. When Schirmer awoke, he realized that twenty minutes were missing in his life. Later, he was put under hypnosis

The next thing Schirmer remembered, under hypnosis, was rolling down the car window and talking to the occupant of the craft, who pressed something against the side of his neck and asked him: "Are you the watchman over this place?" He then pointed to a powerplant that was visible and asked him, "Is this the only source of power you have?"

Schirmer was taken aboard the craft. He saw control panels and computer-like machines. The occupants appeared to be normal human beings and were wearing coveralls with an emblem of a winged serpent. They told him their craft operated by reverse electromagnetism and they drew their power from large water reservoirs.

"To a certain extent they want to puzzle people," Schirmer reported under hypnosis. At one point, one of the men took Schirmer to the large window of the ship, pointed to the deserted landscape around them and said gravely, "Watchman, someday you will see the Universe!" In an apparent attempt at disinformation, they told him they were from another galaxy, that he would not remember being inside the ship, and concluded: "You will not speak wisely about this night. We will return to see you two more times."[25]

In a similar incident near Temple, Oklahoma on March 23, 1966, an instructor in aircraft electronics at Sheppard Air Force Base was driving to work at 5:00 A.M. on Route 65, approaching the intersection with Highway 70. He suddenly encountered a large fish-shaped object blocking the highway. He pulled over and approached the object, seeing a man run inside the strange craft, which then rose rapidly.

In a telephone interview, the instructor told UFO researcher Jacques Vallee, "One mile before the intersection I saw a very bright light a mile or so to my right, and I supposed it was a truck having trouble on the highway. I went on to turn west on Highway 70. I went a quarter of a mile or so, and changed my mind and thought that it was a house that was being moved down the highway in the early morning hours.

"...It was parked on the highway and I got within a hundred yards of it and stopped, got out of the car, and started trotting towards the object, leaving the car lights on and my engine running. I got about fifteen steps or so, and I happened to think I had a Kodak on the front seat, and I would like to get a picture. I hesitated just a second, and while I did, why this man that was dressed in military fatigues, which I thought was a master sergeant... this insignia was on his right arm, and he had a kind of cap with the bill turned up, weighed approximately 180 pounds and about 5'9" ...."

"He looked perfectly ordinary?" asked Jacques Vallee.

"Oh, yes, he was just a plain old G.I. mechanic ... or a crew chief or whatever he might happen to be on that crew. He had a flashlight in his hand, and he was almost kneeling on his right knee, with his left hand touching the bottom of the fuselage."

The object looked like an aluminum airliner with no wings or tail and with no seams along the fuselage. It lifted up vertically for about fifty feet and headed southeast almost straight backward, off by about ten degrees, at a speed estimated to be about 720 mph, judging by the barns it illuminated along its path across the valley. It was the size of a cargo plane, but had no visible means of propulsion. The witness was grilled by a roomful of officers at the Air Force base. A truck driver down the road had observed the same object.

Concludes Vallee on this sighting, "Whoever he was, the man in the baseball cap was no interplanetary explorer. This is only one of many sightings in which the pilots are described as ordinary humans. Whatever they are, the occupants of such craft are not genuine extraterrestrials."[25]

Similarly, on January 25, 1967, an unnamed Minnesota man was driving his pickup truck at 4:30 a.m. near Winsted, Minnesota, when he suddenly stalled. He got out of his truck to see what had happened when a brilliantly illuminated object became visible approaching him. He dashed back into the vehicle and locked the doors. The UFO landed on the road, and a man dressed in blue coveralls and wearing "something like a glass fishbowl on his head" emerged briefly before reentering the craft and flying away.[24]

What is fascinating about these and many other encounters is that, aside from the "fact" that each involves a craft of a completely different design from those we are accustomed to seeing in this century, each of these events appears to be a highly mundane encounter with perfectly normal human beings. The occupants appear to be of our own planet and are not threatening, and often not even apparently observing the witnesses or wanting contact with them. They are merely going about their own business (whatever that may be) or repairing some malfunction on their craft. Some of these "occupants" do have more interaction with others, including the inevitable falling in love with someone from another time.

In the case of the Ashland, Nebraska sighting, the occupants may have been looking for a source of power or something to refuel their ship. Upon seeing the electrical generating station they were prompted to ask (perhaps in surprise at its primitiveness), "Is this the only source of power you have?" Later when Schirmer is informed that 'one day he will see the stars,' this may merely mean that time travel and space travel is within the easy reach of all mankind.

The Time Travel Prayer

Imagine there is a bank that credits your account each morning with $86,400. It carries over no balance from day to day. Every evening it deletes whatever part of the balance you failed to use during the day. What would you do? Draw out every penny, of course!

Each of us has such a bank. Its name is TIME.

Every morning, it credits you with 86,400 seconds. Every night it writes off, as lost, whatever of this you have failed to invest to good purpose.

It carries over no balance. It allows no overdraft. Each day it opens a new account for you. Each night it burns the remains of the day. If you fail to use the day's deposits, the loss is yours. There is no going back. There is no drawing against the "tomorrow." You must live in the present on today's deposits. Invest it so as to get from it the utmost in health, happiness, and success! The clock is running. Make the most of today.

To realize the value of ONE YEAR, ask a student who failed a grade.
To realize the value of ONE MONTH, ask a mother who gave birth to a premature baby.
To realize the value of ONE WEEK, ask the editor of a weekly newspaper.
To realize the value of ONE HOUR, ask the lovers who are waiting to meet.
To realize the value of ONE MINUTE, ask a person who missed the train.

To realize the value of ONE SECOND, ask a person who just avoided an accident.
To realize the value of ONE MILLISECOND, ask the person who won a silver medal in the Olympics.

Treasure every moment that you have! And treasure it more because you shared it with someone special—special enough to spend your time on.

And remember that time waits for no one. Yesterday is history. Tomorrow is a mystery. Today is a gift. That's why it's called the present!

Above: The previously missing photo taken by Dr. S. Milliken on March 15, 1901, of a mysterious airship over Silver City, New Mexico. It apparently shows three cigar shaped objects lashed together. Time travel airships from the future? Photos courtesy of Dr. Milliken's great grandson Ed Phipps. Below: A close-up of the unidentified craft.

An illustration of the airship seen over Chicago in 1896.

From the late 1800s to now (and into the future) there are many strange connections and coincidences concerning time travel, starting with H.G. Wells' 1895 book *The Time Machine*.

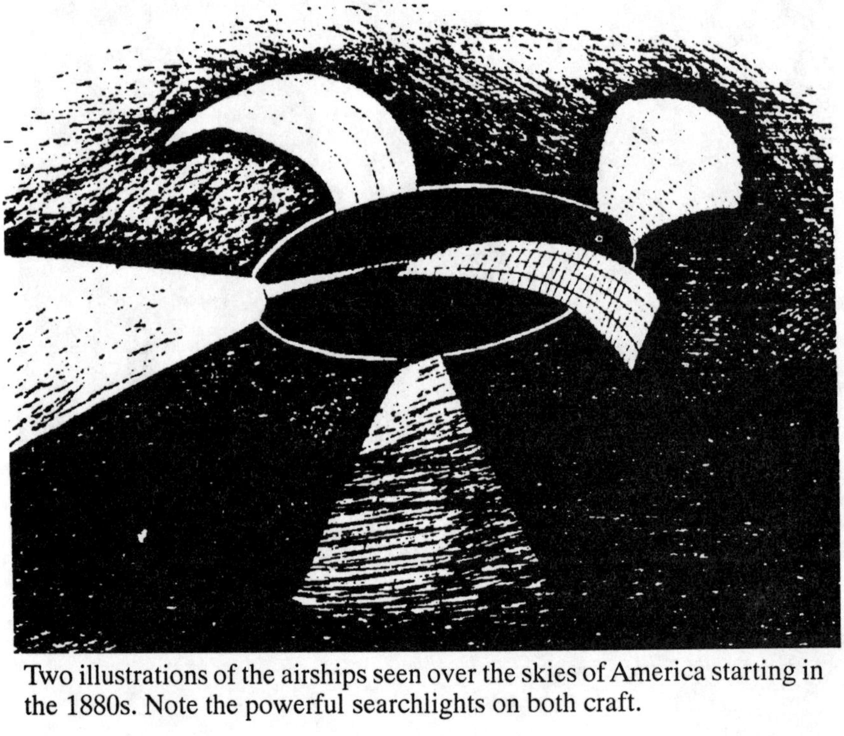

Two illustrations of the airships seen over the skies of America starting in the 1880s. Note the powerful searchlights on both craft.

Above: On August 20, 1949, Professor Clyde Tombaugh, the astronomer credited with discovering Pluto in 1930, saw this giant cigar-shaped UFO outside of Las Cruces, New Mexico. It was very similar to the craft described in the late 1800s.

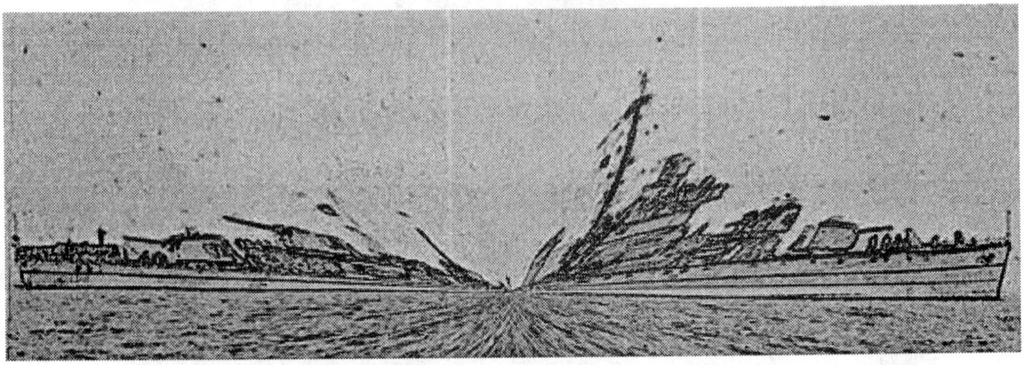

Above: Was a U.S. Navy battleship sent through time and space as theorized in the book *The Philadelphia Experiment?* Below: A cover from *Fate* magazine, the lead story being Al Bielek's account of Project Rainbow.

4.
The Philadelphia Experiment A History

Actually, the biggest deterrent to scientific progress is a refusal of some people, including scientists, to believe that things that seem amazing can really happen.
—George S. Trimble, director of the NASA Manned Spacecraft Center, Houston

The complete time traveller should be in-the-know on important time travel facts in general, including the story of the Philadelphia Experiment. The implications of the Philadelphia Experiment are far reaching, and the whole episode should be seen as a turning point in history, as far as time travel is concerned.

Where it all Began

The "Philadelphia story" starts with Dr. Morris K. Jessup. Dr. Jessup was an astronomer, an author, and the highly active president of a UFO club.

Morris K. Jessup was born on March 20, 1900 in Rockville, Indiana. Jessup served in WWI and later studied mathematics and astronomy at Drake University, Des Moines, Iowa, and later at the University of Michigan in Ann Arbor. As a doctoral student in the late 1920s, he travelled to South Africa and worked at the Lamont-Hussey Observatory in Bloemfontein.

Back in Michigan, Jessup earned a doctorate in astrophysics in 1933. During these Depression years there were few jobs for astronomers or astrophysicists, so he worked for the U.S. Department of Agriculture as part of team of scientists going to Brazil to "study the sources of crude rubber in the headwaters of the Amazon."[7] A strange assignment indeed, since the headwaters of the Amazon actually would be situated in Peru,

Bolivia or Ecuador.

Jessup, during these years, and on subsequent journeys to Mexico and Peru, visited the massive ruins at Teotihuacan, and at Cuzco, Sacsayhuaman, Ollantaytambo and Machu Picchu. Jessup concluded that these giant walls could not have been built by the Incas, as archaeologists commonly suggest. Rather, he formulated one of the earliest "Gods from Outer Space" theories, and went on to have a book which included the theory published in 1955. This book was entitled *The Case for the UFO* and it was published by Citadel Press in New York, a major publisher of books on the occult at the time. The next year, Jessup brought out two quick books for Citadel, *The UFO Annual* and *UFOs and the Bible.*

Jessup's life was suddenly to involve a time traveller who called himself several names, among them Carlos Miguel Allende and Carl M. Allen. Allende/Allen had written to Jessup in 1955 when *The Case For the UFO* had first come out. Jessup had written a quick note in return, and continued to work on his other books for Citadel and lecture around the U.S. about UFOs and Einstein's Unified Field Theory.

Jessup would tell audiences that, "If the money, thought, time, and energy now being poured uselessly into the development of rocket propulsion were invested in a basic study of gravity, beginning perhaps with continued research into Dr. Einstein's Unified Field concepts, it is altogether likely that we could have effective and economical space travel, at but a fraction of the costs were are now incurring, within the next decade."[7]

Allende/Allen attended at least one of Jessup's talks during 1955/56 and wrote Jessup a second letter. It was postmarked from Miami, on the letterhead of the Turner Hotel in Gainesville, Texas, and carried a Pennsylvania rural address for return.

In these letters, we are given the first clues about the Philadelphia Experiment and what actually happened back in 1943.

The Allende Letters

Allende's letters addressed to Dr. Jessup are important letters in reconstructing what actually went on in the government's early time travel and teleportation experiments. Note that in the following paragraphs, Allende's peculiar punctuation and syntax are unchanged. His use of all capitals and odd writing style are left intact. Following are some excerpts from his long letters:

Carlos Miguel Allende
R.D. #1, Box 223
New Kensington, Penn.

My Dear Dr. Jessup,

Your invocation to the Public that they move en Masse upon their Representatives and have thusly enough Pressure placed at the right & sufficient Number of Places where from a Law demanding research into Dr. Albert Einstein's Unified Field Theory May be enacted (1925-27) is Not at all Necessary. It May Interest you to know that The Good Doctor Was Not so Much influenced in his retraction of that Work, by Mathematics, as he most assuredly was by Humantics.

His Later computations, done strictly for his own edification & amusement, upon cycles of Human Civilization & Progress compared to the Growth of Mans General over-all Character Was enough to Horrify Him. Thus, We are "told" today that that Theory was "Incomplete."

Dr. B. Russell asserts privately that It is complete. He also says that Man is Not Ready for it & Shan't be until after W.W. III. Nevertheless, "Results" of My friend Dr. Franklin Reno, Were used. These Were a complete Recheck of That Theory, With a View to any & Every Possible quick use of it, if leasable in a Very short time. There Were good Results, as far as a Group Math Re-Check AND as far as a good Physical "Result," to Boot. YET, THE NAVY FEARS TO USE THIS RESULT. The Result was & stands today as Proof that The Unified Field Theory to a certain extent is correct. Beyond that certain extent No Person in his right senses, or having any senses at all, Will evermore dare to go. I am sorry that I have Mislead You in My Previous Missive. True, enough, such a form of Levitation has been accomplished as described. It is also a Very commonly observed reaction of certain Metals to Certain Fields surrounding a current. This field being used for that purpose. Had Farraday concerned himself about the Mag. field surrounding an electric Current, We today Would NOT exist or if We did exist, our present Geo-political situation would have the very time-bomish, ticking off towards Destruction, atmosphere that Now exists. Alright, Alright! The "result" was complete invisibility of a ship, Destroyer type, and all of its crew, While at Sea. (Oct. 1943) The Field Was effective in an oblate spheroidal shape, extending one hundred yards (More or Less, due to Lunar position & Latitude) out from each beam of the ship. Any Person Within that sphere became vague in form BUT He too observed those Persons aboard that ship as though they too were of the same state, yet were walking upon nothing. Any person without that sphere could see Nothing save the clearly Defined shape of the Ships Hull in the Water, PROVIDING of course, that that person was just close enough to see, yet, just barely outside of that field. Why tell you Now? Very Simple; If You choose to go Mad, then you would reveal this information. Half of the officers & the crew of that Ship are at present, Mad as Hatters. Afew, are even Yet, confined to certain areas where they May receive trained Scientific aid when they, either, "Go Blank" or "Go Blank" & Get Stuck." Going-Blank IE an after effect of the Man having been within the field too Much, IS Not at all an unpleasant experience to Healthily Curious Sailors. However it is when also, they "Get Stuck" that they call it "HELL" INCORPORATED" The Man thusly stricken can Not Move of his own volition unless two or More of those who are within the field go & touch him, quickly, else he "Freezes".

If a Man Freezes, His position Must be Marked out carefully and then the field is cut-off. Everyone but that "Frozen" Man is able to Move; to appreciate apparent Solidity again. Then, the Newest Member of the crew Must approach the Spot, where he will find the "Frozen" Mans face or Bare Skin, that is Not covered by usual uniform Clothing. Sometimes, It takes only an hour or so Sometimes all Night & all Day Long & Worse It once took 6 months, to get The Man "Unfrozen". This "Deep Freeze" was not psychological. It is the Result of a Hyper-Field that is set up, within the field of the Body. While The "Scorch" Field is turned on & this at Length *or* upon a Old Hand.

Ahighly complicated Piece of Equipment Had to be constructed in order to Unfreeze those who became "True Froze" or "Deep Freeze" subjects. Usually a "Deep Freeze" Man goes Mad, Stark Raving, Gibbering, Running Mad, if His "freeze" is far More than a Day in our time.

I speak of TIME for DEEP "Frozen Men" are Not aware of Time as We know it. They are Like Semi-comatose person, who Live, breathe, Look & feel but still are unaware of So Utterly Many things as to constitute a "Nether World" to them. AMan in an ordinary common Freeze *is* aware of Time, Sometimes acutely so. Yet They are Never aware of Time precisely as you or I are aware of it. The First "Deep Freeze" as I said took 6 months to Rectify. It also took over 5 Million Dollars worth of Electronic equipment & a Special Ship Berth.If around or Near the Philadelphia Naval Yard you see a bunch of Sailors in the act of Putting their Hands upon a fellow or upon "thin air", observe the Digits & appendages of the Stricken Man. If they seem to Waver, as tho witihin a Heat-Mirage, go quickly & Put YOUR Hands upon Him, For that Man is The Very Most

Desperate of Men in the World. No one of those Men ever want at all to become again invisible. I do Not think that Much More Need be said as to Why Man is Not Ready of Force-Field Work. Eh?"

"This 'Deep Freeze' was not psychological. It is the result of a Hyper-Field that is set up, within the field of the body, while the 'Scorch' Field is turned on..."

"The Navy did not know that the men could become invisible WHILE NOT UPON THE SHIP & UNDER THE FIELD'S INFLUENCE... The Navy did not know that there would be men die from odd effects of HYPER 'Field' within or upon 'Field'... Further, they even yet do not know why this happened and are not even sure that the 'F' within 'F' is the reason, for sure at all."

"There was plenty of static electricity associated with it."

"I actually shoved my hand, up to the elbow, into this unique force field as that field flowed, surging powerfully in a COUNTERclockwise direction around the little experimental Navy ship, the DE 173. I felt the...push of that force field against the solidness of my arm and hand outstretched into its humming-pushing-propelling flow.

"I watched the air all around the ship...turn slightly, ever so slightly, darker than all the other air...I saw, after a few minutes, a foggy green mist arise like a thin cloud. ... this must have been a mist of atomic particles. I watched as thereafter the DE 173 became rapidly invisible to human eyes. And yet, the precise shape of the keel and underhull of that...ship REMAINED impressed into the ocean water as it and my own ship sped along somewhat side by side and close to inboards...

"...in trying to describe the sounds that [the] force field made as it circled around the DE 173...it began as a humming sound, quickly built up... to a humming whispering sound, and then increased to a strongly sizzling buzz—[like a] rushing torrent...

"The field had a SHEET of pure electricity around it as it flowed. [This] flow was strong enough to almost knock me completely off balance and had my entire body been within that field, the flow would of a most absolute certainty [have] knocked me flat... on my own ship's deck. As it was, my entire body was NOT within that force field when it reached maximum strength-density, repeat, DENSITY, and so I was not knocked down but my arm and hand [were] only pushed backward with the field's flow.

"Why was I not electrocuted the instant my bare hand touched that... sheet of electricity surrounding the field flow? It must have been because... [I was wearing] hip-high rubber sailor's boots and sou'wester coat.

"...Naval ONR scientists today do not yet understand what took place that day. They say the field was 'reversed'. Scientific history, I later came to realize, was made for the first time that day."

Project Rainbow

Allende's letters are heady stuff, indeed! One can see how they piqued the interest of time travel researchers. They describe the alleged results of a U.S. Navy time travel experiment now known as the Philadelphia Experiment. Several books, and at least one movie, have been made concerning this fascinating story. Sometimes truth is stranger than fiction, but sifting the truth from the fiction in terms of the U.S. military and their alleged time travel experiments (as well as anti-gravity experiments, flying saucer building and possible extraterrestrial collaboration) is a difficult and time consuming task, if you'll pardon the pun.

According to Al Bielek (an alleged survivor of the experiment who will be discussed in detail later), the Philadelphia Experiment had its origin in a feasibility study started in the 1931 at the University of Chicago and later moved to Princeton's Institute of Advanced Studies. At that time the project was known as "Project Invisibility," and involved in the program at various times were renowned scientists that included Dr. Albert Einstein, Dr. John von Neumann, Dr. Nikola Tesla, a Dr. Alexander, and others.

Bielek claims that he took part in the study as Edward Cameron, a physicist and Naval officer, born on August 4, 1916. His younger brother, Duncan Cameron (born May, 1917), also took part in the experiment.

Originally a theoretical study in methods to produce an "invisibility screen," the first fully successful test in 1940 converted the program to a military project : "Project Rainbow." The project was studied at Princeton, with Nikola Tesla in charge of the time travel experiments and Albert Einstein consulting on the project. Actual tests of theories were to be undertaken at the Philadelphia Naval Yard starting in January of 1942, hence, "The Philadelphia Experiment."

William Moore mentions in his book *The Philadelphia Experiment* that when he contacted the Navy about any records concerning a secret undertaking code named "Project Rainbow," he was told by the Navy that no such designation as "Rainbow" existed within the Navy records.

However, when Moore checked the National Archives and obtained The Inter-Services Code-Word Index (issued September 1, 1941), the code-word "Rainbow" was listed with a block number designation of 334, thereby proving the Navy absolutely wrong in their denial of any code-word designation concerning the word.

According to Bielek, as a military (Navy) program, a volunteer crew had to be trained and a ship selected—the *Eldridge* (DE 173). Equipment had to be ordered, built, and installed for seaborne tests. Sea trials were conducted outside of Philadelphia, in the Delaware Bay area.

Bielek says that it was Tesla's intention to turn a battleship invisible, but he deliberately sabotaged the January 1942 test, as he had decided that they were not yet ready to try the dangerous experiment. At this point, Tesla was replaced by Dr. John von Neumann. Tesla had designed an analog system of invisibility, but Dr. von Neumann replaced it with a "digital-pulsed system." This digital-pulsed system was later to be used in the Montauk Experiments, according to Bielek and Preston Nichols.

Four large coils of of single-turn copper tubing were used to wind parts of the deck, and an antenna array was situated at the top of the highest mast of the *USS Eldridge.* It is interesting to note that during this year, 1943, no copper penny was issued. A steel penny was issued instead. Why?—Because all of the copper was being used in the copper cables used to wind the *Eldridge,* and other ships, for degaussing and invisibility experiments!

The first test was on July 22, 1943 with serious personnel problems resulting. Four coils radiating electromagnetic power were used to create the field around the ship. Invisibility was achieved, according to Bielek, but sailors on the ship got nauseous and reported an uncomfortable "sick" feeling.

The final test—12 August 1943—was an utter disaster to ship and crew. According to Bielek there was a sudden blue flash of light and the ship disappeared. The sailors on the ship, which included Edward Cameron and his brother Duncan, saw normal space-time disintegrate before their very eyes and they entered another dimension. The radios on board the ship did not work. The ship survived after its return to normal time and location in the Philadelphia harbor, but many of the crew were either insane, missing, or even fused with portions of the steel ship in the most gruesome and horrifying manner.

The U.S. Navy continues to deny to this day that this project ever took place, while other authors (such as Preston Nichols) claim that Project Rainbow was coupled through hyperspace to another project in the future, the 1983 Montauk Project.

But first, let us start with the official Navy response to enquiries concerning these invisibility, teleportation and time travel experiments.

The Official Navy Press Release Concerning the Philadelphia Experiment

Starting in 1979 with the publication of William Moore and Charles Berlitz's bestselling book, *The Philadelphia Experiment,* the whole subject of time travel and teleportation was thrust into the public arena. The film version appeared in 1984, and continued reference to the Philadelphia Experiment happened regularly.

The U.S. Navy was suddenly deluged with requests for information on the alleged experiment, Project Rainbow and the *USS Eldridge.* So many requests were made to the Navy, that they created an official response, one that can also be found on the world-wide web if one searches for the words "Philadelphia Experiment."

What follows is the official response drafted by the Navy:

DEPARTMENT OF THE NAVY—NAVAL HISTORICAL CENTER
901 M STREET SE—WASHINGTON NAVY YARD
WASHINGTON DC 20374-5060

The "Philadelphia Experiment"

Allegedly, in the fall of 1943 a U.S. Navy destroyer was made invisible and teleported from Philadelphia, Pennsylvania, to Norfolk, Virginia, in an incident known as the Philadelphia Experiment. Records in the Operational Archives Branch of the Naval Historical Center have been repeatedly searched, but no documents have been located which confirm the event, or any interest by the Navy in attempting such an achievement.

The ship involved in the experiment was supposedly the *USS Eldridge.*

Operational Archives has reviewed the deck log and war diary from *Eldridge*'s commissioning on 27 August 1943 at the New York Navy Yard through December 1943. The following description of *Eldridge*'s activities are summarized from the ship's war diary. After commissioning, *Eldridge* remained in New York and in the Long Island Sound until 16 September when it sailed to Bermuda. From 18 September, the ship was in the vicinity of Bermuda undergoing training and sea trials until 15 October when *Eldridge* left in a convoy for New York where the convoy entered on 18 October. *Eldridge* remained in New York harbor until 1 November when it was part of the escort for Convoy UGS-23 (New York Section). On 2 November the convoy entered Naval Operating Base, Norfolk. On 3 November, *Eldridge* and Convoy UGS-23 left for Casablanca where it arrived on 22 November. On 29 November, *Eldridge* left as one of the escorts for Convoy GUS-22 and arrived with the convoy on 17 December at New York harbor. *Eldridge* remained in New York on availability training and in Block Island Sound until 31 December when it steamed to Norfolk with four other ships. During this time frame, *Eldridge* was never in Philadelphia.

Eldridge's complete World War II action report and war diary coverage, including the remarks section of the 1943 deck log, is available on microfilm, NRS-1978-26. The cost of a duplicate film is indicated on the fee schedule. To order a duplicate film, please complete the duplication order form and send a check or money order for the correct amount as indicated on the NHC fee schedule, made payable to the Department of the navy, to the Operational Archives, at the above address.

Supposedly, the crew of the civilian merchant ship *SS Andrew Furuseth* observed the arrival via teleportation of the *Eldridge* into the Norfolk area. *Andrew Furuseth*'s movement report cards are in the Tenth Fleet records transferred to the Textual Reference Branch, National Archives and Records Administration, 8601 Adelphi Road, College Park, MD 20740-6001. The cards list the ship's ports of call, the dates of the visit, and convoy designation, if any. The movement report card shows that *Andrew Furuseth* left Norfolk with Convoy UGS-15 on 16 August 1943 and arrived at Casablanca on 2 September. The ship left Casablanca on 19 September and arrived off Cape Henry on 4 October. *Andrew Furuseth* left Norfolk with Convoy UGS-22 on 25 October and arrived at Oran on 12 November. The ship remained in the Mediterranean until it returned with Convoy GUS-25 to Hampton Roads on 17 January 1944. The Archives has a letter from Lieutenant Junior Grade William S. Dodge, USNR, (Ret.), the master of *Andrew Furuseth* in 1943, categorically denying that he or his crew observed any unusual event while in Norfolk. *Eldridge* and *Andrew Furuseth* were not even in Norfolk at the same time.

The Office of Naval Research (ONR) has stated that the use of force fields to make a ship and her crew invisible does not conform to known physical laws. ONR also claims that Dr. Albert Einstein's Unified Field Theory was never completed. During 1943-1944, Einstein was a part-time consultant with the Navy's Bureau of Ordnance, undertaking theoretical research on explosives and explosions. There is no indication that Einstein was involved in research relevant to invisibility or to teleportation. ONR's information sheet on the Philadelphia Experiment is attached.

The Philadelphia Experiment has also been called "Project Rainbow." A comprehensive search of the Archives has failed to identify records of a Project Rainbow relating to teleportation or making a ship disappear. In the 1940s, the code name RAINBOW was used to refer to the Rome-Berlin-Tokyo Axis. The RAINBOW plans were the war plans to defeat Italy, Germany and Japan.

RAINBOW V, the plan in effect on 7 December 1941 when Japan attacked Pearl Harbor, was the plan the U.S. used to fight the Axis powers.

Some researchers have erroneously concluded that degaussing has a connection with making an object invisible. Degaussing is a process in which a system of electrical cables are installed around the circumference of ship's hull, running from bow to stern on both sides. Ameasured electrical current is passed through these cables to cancel out the ship's magnetic field. Degaussing equipment was installed in the hull of Navy ships and could be turned on whenever the ship was in waters that might contain magnetic mines, usually shallow waters in combat areas. It could be said that degaussing, correctly done, makes a ship "invisible" to the sensors of magnetic mines, but the ship remains visible to the human eye, radar, and underwater listening devices.

After many years of searching, the staff of the Operational Archives and independent researchers have not located any official documents that support the assertion that an invisibility or teleportation experiment involving a Navy ship occurred at Philadelphia or any other location.

11 December 1998

End of Official Navy Press Release

The Official Office of Naval Research's Response to the Philadelphia Experiment

DEPARTMENT OF THE NAVY—NAVAL HISTORICAL CENTER
901 M STREET SE—WASHINGTON NAVY YARD
WASHINGTON DC 20374-5060
Related resources: Philadelphia Experiment
DEPARTMENT OF THE NAVY OFFICE OF NAVAL RESEARCH
ARLINGTON, VIRGINIA 22217

Information Sheet: Philadelphia Experiment

Over the years, the Navy has received innumerable queries about the so-called "Philadelphia Experiment" or "Project" and the alleged role of the Office of Naval Research (ONR) in it. The majority of these inquiries are directed to the Office of Naval Research or to the Fourth Naval District in Philadelphia. The frequency of these queries predictably intensifies each time the experiment is mentioned by the popular press, often in a science fiction book.

The genesis of the Philadelphia Experiment myth dates back to 1955 with the publication of *The Case for UFOs* [sic] by the late Morris K. Jessup.

Some time after the publication of the book, Jessup received correspondence from a Carlos Miquel Allende, who gave his address as R.D. #1, Box 223, New Kensington, Pa. In his correspondence, Allende commented on Jessup's book and gave details of an alleged secret naval experiment conducted by the Navy in Philadelphia in 1943. During the experiment, according to Allende, a ship was rendered invisible and teleported to and from Norfolk in a few minutes, with some terrible after-effects for crew members.

Supposedly, this incredible feat was accomplished by applying Einstein's "unified field" theory. Allende claimed that he had witnessed the experiment from another ship and that the incident was reported in a Philadelphia newspaper. The identity of the newspaper has never been established. Similarly, the identity of Allende is unknown, and no information exists on his present address.

In 1956 a copy of Jessup's book was mailed anonymously to ONR. The pages of the book were interspersed with hand-written comments which alleged a knowledge of UFO's, their means of motion, the culture and ethos of the beings occupying these UFO's, described in pseudo-scientific and incoherent terms.

Two officers, then assigned to ONR, took a personal interest in the book and showed it to Jessup. Jessup concluded that the writer of those comments on his book was the same person who had written him about the Philadelphia Experiment. These two officers personally had the book retyped and arranged for the reprint, in typewritten form, of 25 copies. The officers and their personal belongings have left ONR many years ago, and ONR does not have a file copy of the annotated book.

Personnel at the Fourth Naval District believe that the questions surrounding the so-called "Philadelphia Experiment" arise from quite routine research which occurred during World War II at the Philadelphia Naval Shipyard. Until recently, it was believed that the foundation for the apocryphal stories arose from degaussing experiments which have the effect of making a ship undetectable or "invisible" to magnetic mines. Another likely genesis of the bizarre stories about levitation, teleportation and effects on human crew members might be attributed to experiments with the generating plant of a destroyer, the *USS Timmerman*. In the 1950's this ship was part of an experiment to test the effects of a small, high-frequency generator providing 1,000 hz instead of the standard 400hz. The higher frequency generator produced corona discharges, and other well known phenomena associated with high frequency generators.

None of the crew suffered effects from the experiment.

ONR has never conducted any investigations on invisibility, either in 1943 or at any other time (ONR was established in 1946.) In view of present scientific knowledge, ONR scientists do not believe that such an experiment could be possible except in the realm of science fiction.

08 September 1996
End Report

According to the official Navy line, Project Rainbow had to do with plans to defeat the Axis, and invisibility experiments were never carried out. The descriptions of the experiments they do admit to are interesting, however! Also interesting is the Office of Naval Research's reference to its handling of a copy of Jessup's book *The Case for the UFO* received from an anonymous sender. Other sources say there is more to this story than has been related.

The Varo Edition

Riley Crabb, the former director of the Borderland Research Sciences Foundation in California, was one of the first people to write about the Philadelphia Experiment. He published a mimeographed booklet in 1962 entitled *M.K. Jessup, the Allende Letters and Gravity.*[41] According to Crabb, an annotated copy of *The Case for the UFO* was addressed to Admiral N. Furth, Chief, Office of Naval Research, Washington 25, D.C., and was mailed in a manila envelope postmarked Seminole, Texas, in 1955. In July or August of that year the book appeared in the incoming correspondence of Major Darrell L. Ritter, USMC, Aeronautical Project Office in ONR. When Captain Sidney Sherby, a newcomer, reported aboard at ONR he obtained the book from Major Ritter. Captain Sherby and Commander George W. Hoover, Special Projects Officer, ONR, indicated interest in some of the notations the book contained.

The paperback copy of Jessup's book was marked throughout with underlinings and notations, evidently by three different persons, for three distinct colors of ink were used—blue, blue-violet, and blue-green. The notations in Jessup's book implied intimate knowledge of flying saucers, their means of propulsion, their origin, background, history, and even the habits of the beings occupying them. In the annotated bits in the margins of the book were references and unusual terms such as mothership, home ship, dead ship, great ark, great bombardment, great return, great war, little-men, force fields, deep freezes, measure markers, scout ships, magnetic fields, gravity fields, sheets of diamond, cosmic rays, force cutters, inlay work, clear talk, telepathing, nodes, vortices, magnetic net, and other strange references.

Hoover and Sherby invited Jessup to visit them at the Office of Naval Research in Washington in the spring of 1957 to discuss the book they had received. They handed Jessup the annotated copy and asked him to examine the handwritten notes.

"This book was sent to us through the mail anonymously. Apparently it was passed back and forth among at least three persons who made notations. Look it over, Mr. Jessup, and tell us if you have any idea who wrote those comments," one of the officers said.

As Jessup looked at the book, he was troubled at what he saw. Due to the uniqueness of Allende's spelling and grammar, it became almost immediately evident to Jessup that the letter-writer Allende was possibly one of those who had worked on the annotation of his book, and was very likely the same individual who had originally mailed the volume to the Navy in the first place. The many references were startling and showed that the writer(s) were extremely knowledgeable in the subject of UFOs, folklore, mysticism and physics. Further troubling was the fact that Navy took such an interest in these strange ramblings. Jessup finally told them that he had received two letters from one of the writers in the book, Carlos Allende / Allen.

"Thank you, Mr. Jessup," he was told. " It is important that we see those letters." Hoover then told Jessup that he was arranging for the printing of a special edition of the book "for some of our top people," and that he would be sure to arrange for Jessup to get a copy.

Jessup returned to his home and supplied Hoover and Sherby with copies of the letters. The two officers then arranged for the Varo Manufacturing Company of Garland, Texas to produce a mimeographed edition of the book. Brad Steiger, a well-known author on occult and paranormal issues, claims that the Varo company is engaged in "secret government work."[9] This special edition of Jessup's book, with the handwritten notes in the margins, became known as "the Varo annotated edition," or just "the Varo edition."

It is interesting to note that the Navy would not ordinarily pay attention to such a thing, but they did take note of this submission and what was written in the margins. Why would they have paid any attention to someone whose name they should not recognize unless they were perhaps affiliated with the Philadelphia Experiment?

Since the book allegedly involved the military and their apparent knowledge of technical details concerning UFOs, the Varo edition of *The Case for the UFO* caused considerable interest within the fringe community of researchers who were studying the UFO enigma.

Brad Steiger says that Riley Crabb sent correspondence to him on September 24, 1962, where Crabb clears up the mystery of how he knew of the Varo edition's history, and how he happened to obtain a copy of the original Varo edition. It was a copy that the Navy originally gave to Jessup, and apparently was given to Crabb by Jessup. This copy, however, rather mysteriously disappeared in April, 1960, when Crabb mailed it to himself from Washington.

In elaborating further about the original Varo edition, Crabb wrote to Steiger, "It may be that CDR Hoover, or some other open minded officer in the Office of Navy Research asked Varo to print the notes and the book; on the other hand, maybe Bob Jordan or some other official at Varo, in Washington at the time, sniffed a possible research and development project, and volunteered to do it. Anyhow, I understand that 25 copies were reproduced on standard letter paper, 8 1/2 x 11, probably on Varo's own little litho press, and plastic bound, pretty close to 200 pages. Michael Ann Dunn, the stenographer who did the editing, explains why in the introduction. She's married now, living in Dallas, and won't answer her phone. Garland (where Varo is located) is a Dallas suburb. I suppose Varo and/or the Navy judiciously handed out a few copies to those capable of picking up a hot lead on the anti-gravity research trail. From their comment, Allende and his gypsy friends were not educated, not technicians, and did not give any illustrations or mathematical formulae which would help build usable hardware.

"Varo, by the way, is a small manufacturing firm in electronics and up to its neck in space age business. Apparently it has succeeded in developing some kind of a death ray gadget, judging from a guarded press release of last fall when a group of Congressmen visited there for a demonstration. I think it extremely interesting that certain Naval officers and Varo officials took the Allende notes far more seriously at first reading than did Jessup himself! My first reaction to them was one of skepticism, but now I believe them. So does one of our most material minded, hardest headed electronic engineer associates in Los Angeles."[9]

Crabb, in *M.K. Jessup, the Allende Letters and Gravity,* reprints Michael Ann Dunn's Introduction to the Varo edition. She identifies three different people as making the notations, each in different colors of ink, and refers to them as "Mr. A," "Mr. B," and "Jemi." It was assumed that the third person was "Jemi" because of the direct use of that name in salutations and references by Mr. A and Mr. B throughout the book.

Dunn speculates on the possibility that two of the persons are twins, since there are two references to this word, appearing on pages six and 81 of the original. "The assumption that Mr. A is one of the twins may be correct. On page 81, Mr. A has written and then marked through, '...and I Do Not Know How this came to Pass, Jemi.' Then he has written, 'I remember My twin...' On page six he writes in an apparent answer to Mr. B, "No, My twin...' We cannot be sure of the other twin."

"It is probable," Ms. Dunn continues, "that these men are Gypsies. In the closing pages of the book Mr. B says, '...only a Gypsy will tell another of that catastrophe. And we are a discredited people, ages ago. Hah! Yet, man wonders where "we" come from...' On page 130 Mr. A says, '...ours is a way of life, time proven & happy. We have nothing, own nothing except our music & philosophy & are happy.' On page 76 Mr. B says, 'Show this to a Brother Gypsy...' On page 158 the reference to the word 'we' by Mr. A could refer to the 'discredited people.' Charles G. Leland in his book, *English Gypsies and Their Language,* states that the Gypsies call each other brother and sister, and are not in the habit of admitting to their fellowship people of a different blood and with whom they have no sympathy. This could explain the usage of the term in the closing notes, 'My Dear Brothers,' and perhaps the repeated reference to 'vain humankind'."[41]

The paperback copy of *The Case for the UFO* apparently passed through the hands of these men several times. Says Dunn, "This conclusion is drawn from the fact that there are discussions between two or all three of the men, questions, answers, and places where parts of a note have been marked through, underlined, or added to by one or both of the other men. Some have been deleted by marking through."[41]

The Varo edition of Jessup's book may have alarmed certain commanders at the Office of Naval Research. Perhaps they thought that a very serious security leak was occurring, and Jessup had found out too much concerning the Navy's secret time travel and teleportation experiments.

Murder: Men In Black Style

Sadly, the Allende letters and Varo edition of Jessup's book were the beginning of the end for him.

Gray Barker reports in his 1963 mimeographed book *The Strange Case of Dr. M.K. Jessup* that he first learned of the annotated copy when he was talking to Mrs. Walton Colcord John, director of the *Little Listening Post,* a UFO and New Age publication in Washington D.C. sometime in the early 1960s. Speaking over the telephone, Mrs. John told Steiger of a strange rumor going around, to the effect that somebody had sent a marked-up copy to Washington, and that the government had gone to the expense of mimeographing the entire book, so that all the underlinings and notations could be added to the original text. This was being circulated rather widely, she told him, through military channels.

In late October of 1958 Jessup travelled from Indiana to New York, and sometime around Halloween, Jessup contacted Ivan T. Sanderson, the founder of the Society for the Investigation of the Unexplained (SITU). During this meeting, Jessup told Sanderson his bizarre story, and gave a copy of the Varo edition to him, one of three given to him by ONR.

Jessup also confided in Sanderson things of a confidential nature, probably about the Philadelphia Experiment and time travel. Jessup also apparently thought he was being followed or at least his mail was being tampered with.

Jessup had apparently made several further trips to the offices of ONR after the printing of the Varo edition. Hoover and Shelby reportedly made efforts to track down Allende/Allen, with Hoover checking out the rural Pennsylvania address, which turned out to be bogus.

Meanwhile, Jessup had disappeared, and his publisher made efforts to contact him. Eventually, it was discovered that Jessup had driven directly to Florida from New York, where he had intended on living. He was apparently fleeing his home in Indiana, possibly believing he was being watched. He had had a car accident as well, but had survived. He had been in the hospital for an extended period.

Then, on April 20, 1959, he was found dead in his parked car in the rural Dade County Park near his Coral Gables home. A rubber hose ran from his exhaust pipe into the nearly closed back window of the car. His death was ruled as a self-inflicted carbon-monoxide poisoning.

But some, including Ivan T. Sanderson, believed that Jessup had not killed himself, but instead had been "suicided" by the Men in Black who had taken a deep interest in his contact with the time traveller named Carlos Allende and his knowledge of the so-called Philadelphia Experiment.

Ivan T. Sanderson & Ian Fleming

In a curious sidenote, the popular James Bond "007" author, Ian Fleming, is strangely linked to the Philadelphia Experiment. According to Preston B. Nichols and Peter Moon in their book *Pyramids of Montauk* (Sky Books, 1995), Ian Fleming knew certain information about the Rainbow Project, the secret endeavor that was to ultimately lead into what is known today as the Philadelphia Project.

Fleming had worked with Aleister Crowley on the Rainbow Project (acknowledged by the Navy to be a plan to defeat the Axis in WWII), his part being a secret mission to meet with Karl Haushofer of the Nazi party in order to get him to convince Rudolf Hess to defect. Fleming met with Haushofer in Lisbon, Portugal, early in the war and persuaded the influential German occultist to talk with Hess on the behalf of Crowley. Both Haushofer and Hess admired Aleister Crowley a great deal, according to Nichols and Moon.[33]

In August of 1964, say Nichols and Moon, Fleming was planning to fly from his home in Jamaica to New Jersey to meet with Ivan T. Sanderson, a biologist and former British Intelligence agent. As a zoologist, Sanderson had written a number of books, including one on Bigfoot and Yeti, and appeared frequently on radio shows and even Johnny Carson's *Tonight Show.*

Sanderson had become intensely interested in the paranormal, and was a personal friend of Morris K. Jessup, as noted above. Sanderson often talked with his friends about UFOs, the Philadelphia Experiment, cryptozoology and other arcane subjects. Like Fleming, Sanderson was a British expatriate.

According to Nichols and Moon, Sanderson had been corresponding with Fleming, exchanging key information about the Rainbow Project; perhaps the involvement of Crowley, and how it related to the Philadelphia Project. Perhaps Fleming had some inside information as to the secret technology to make battleships invisible, teleport them, and ultimately on UFO technology in general.[33]

In any event, Fleming never made it to his rendezvous with Sanderson. He died of a sudden heart attack on August 12th, 1964 at his home in Jamaica. Nichols and Moon point out that this is the 21st anniversary of the Philadelphia Experiment (August 12, 1943). Was Ian Fleming killed because he knew too much about such things as the Kennedy assassination, the Philadelphia Experiment and the genesis of UFO technology?

Al Bielek and the Philadelphia Experiment

Time travel history was made when a scientist/engineer named Al Bielek (whose general account of the happenings was related above) told a large audience at a 1989 UFO conference in Phoenix, Arizona, that he was a survivor of the Philadelphia Experiment and "Project Rainbow."

According to a Mufon Metroplex presentation on Bielek posted online at http:\\www.in-search-of.com:

"Alfred D. Bielek, was born in August 1916 as Edward Orville Cameron, son of Alexander D. Cameron, Sr. He attended several Universities, graduating with a PhD in Physics (1939). Enlisting in the Navy along with his brother, they eventually became directly involved in the "Philadelphia Experiment." The story of how he—Edward Cameron—was eventually removed from the project and became Alfred D. Bielek is a bizarre story of government brainwashing and manipulation, and destruction of a career. The story of his brother is still more bizarre as well as tragic.

Mr. Bielek is today a retired electronics engineer with 30 years of a consulting career behind him. "

Bielek claimed at the conference that he had survived being teleported through a time warp to the future, as depicted in the 1984 movie *The Philadelphia Experiment,* which experience he remembered only after viewing the movie in 1988. Bielek further claimed that he had been brainwashed by an ultrasecret agency that had created another time travel experiment called The Montauk Project.

According to the Mufon report, Bielek claims the idea for the movie was placed in the hands of Thorn EMI (the British producer of the film *The Philadelphia Experiment*) by a time traveller. According to Bielek, Thorn Instruments, an old English manufacturer of labware from the early 1800s, bought into EMI which had in its archives a story of a disappearing ship with a picture of the time traveller who brought it to them.

In the picture, taken in 1890, were the Wilson brothers of Thorn (who purportedly placed the manuscript in the company's archive vault), Aleister Crowley (infamous occultist) and the "traveller,"a person identified as Preston Nichols. Preston Nichols is a past president of the USPA (United States Psychotronic Association), and has written several books on the Montauk experiments. Supposedly Mr. Nichols was shown the picture by Thorn EMI's Chief Archivist, but was not allowed to take a copy. The picture showed him about 10 years older than he was at the viewing in 1989.

In 1986, Bielek had recalled memories of a site in Montauk, Nichols' main line of study. In Bielek's story, he and his brother (as Duncan and Edward Cameron) time travelled to 1983 Montauk as an unexpected outcome of the 1943 Philadelphia Experiment. According to Bielek, every 20 years on the 12th of August, the magnetic energy peaks (i.e., 1943-1963-1983) to allow a synchronicity. In Montauk, the Camerons were to take part in scientific studies, however, Bielek says his brother lost his "time lock," aged about 1 year per hour and died shortly thereafter. He then said his brother was later reborn. Bielek himself was brainwashed, and treated to an age regression, whereby he returned to infancy and was brought up as Al Bielek.

An interesting bit of information was noted by Dr. Bruce Goldberg in his book *Time Travelers from our Future* which may involve the notion of "synchronicity." Although the dates do not follow exactly, they are very close and may support Bielek's claim. Says Goldberg:

"One rather interesting aspect of the Philadelphia Experiment that is not often reported is what occurred in Philadelphia thirty-nine years before that fateful day in 1943. Late in July of 1904 a ship called the *Mohican* reported experiencing an unusual gray cloud that appeared to attach itself to its hull and created unexplainable electrical effects, such as uncontrollable spinning of compass needles and the magnetization of metal objects on board.

"A series of violent electrical storms were noted in Philadelphia at that very time, along with strange lights moving around the city. Some reports sited a strange ship in the Atlantic, not far from the location of the *USS Eldridge* in 1943. Was this the *USS Eldridge* moving back in time by thirty-nine years?"[40]

Bielek's story is strange, indeed, and has drawn a lot of attention and comment from those interested in time travel. More discussion of his explanation of the technology behind it all will be presented in the next chapter.

The following comments by Vangard Sciences' Jerry W. Decker were posted on the In-Search-Of web site. They are "based on personal studies in a variety of subjects including the Philadelphia Experiment and attendance at [Bielek's January 13, 1990] lecture."

"Mr. Bielek is a most engaging speaker. He presents a wide range of popular references during his talks and weaves quite a web.

"However, the information available prior to the 1983 [sic] movie, *The Philadelphia Experiment* does not concern the prospect of time transport.

"The stated focus of the Philadelphia Experiment project was to "achieve radar invisibility." Mr. Bielek states that his "memories began coming back AFTER seeing the movie in 1988." I got the impression that he incorporates new topics into his storyline as fast as he hears them, claiming that his "memory suddenly came back on that." Another major flaw was the statement that Nikola Tesla was in association with and in charge of the project. This is highly incredible since the experiment occurred in July, 43' and Tesla died on January 7, 1943 in New York City.

"Yet another reference which is difficult to believe is that Gustave LeBon was a consultant and worked with Tesla and Einstein on the project."

In Search of the Truth on the Philadelphia Experiment

As noted, the "Philadelphia story" has drawn a huge amount of attention, and several researchers have tried to follow the trail of tantalizing clues to the end. We reprint two such stories here for your edification.

The following is taken from an Internet posting by Henry Ritson at http:\\ecafe.org/philadelphia/index.htm.

> The 'Philadelphia Experiment' is a compelling piece of fringe-science conspiracy theory, that proposes that experiments into invisibility during the second world war led to warships

disappearing and re-appearing up and down the coast of America, with a good deal of inter-dimensional teleportation thrown in for good measure. Henry Ritson rolls up his sleeves, heads off in search of hard evidence, and finds it has vanished without a trace...

My search begins...

My fascination for the so-called 'Philadelphia Experiment' started with the need for an entertaining academic diversion within Cambridge University library. I was stuck there one glum English Sunday, in search of enlightenment about the effects of the introduction of the Sweet Potato on the culture of highland New Guinea. After several hours of less than fascinating study, it struck me that, whilst surrounded by every single document ever to have been published within the shores of the British Isles (as indeed I was) there was probably some subject considerably more exciting and challenging here that I could be searching for. ... I was, as ever, in the company of Sparky, who was a theoretical Physicist. ... Sparky sent me looking for information about the Philadelphia Experiment. ...

I was enthralled.

This was much more like it. Abandoned or covered up evidence of top secret military research that could change the world as we new it. Kewl! And here was I, surrounded by everything that existed in the public domain. I abandoned the 'Sweet Potato Revolution' in great haste and went a-hunting. So what did we find? The definitive version of the 'Philadelphia Experiment' story was contained in the well-thumbed paperback, *The Philadelphia Experiment* by Charles Berlitz (the man who has made the most money out of the Bermuda Triangle outside of the Bermuda Wreck Retrieval Company). And the story goes like this.

[Here, Ritson recounts the main events of the story, much in the way we have reported above.]

The most extreme version of the story in circulation (as featured in the related 1983 [sic] movie—*The Philadelphia Experiment*) goes as far to suggest that aliens then turned up to check on what the hell we were doing to the magnetic fields and dimensions that they rather assumed we left alone, and took the sailors away to a parallel dimension etc. etc. I suspect that somewhere on fringes of Usenet, someone is also suggesting that the Mafia ordered the aliens to abduct the sailors and take them to the grassy knoll to help Elvis take out JFK.

So here we have a continuum. At one end is pure fact. At the other end is pure lunacy. In the middle are shades of an incredible and highly entertaining story that if it is true, is of historic importance. The question is—at what point on the continuum does the truth lie.

That question prompts another question. How good are the sources?

In search of sources within Berlitz's book, and other documents I have studied since then, there seem to be three main types of primary source. These three are actually rather typical of any conspiracy theory, so maybe I have come across some kind of unified theory after all.

1. Very intense and mysterious letters. ...

2. The 'retired military man who comes out of the woodwork to give tantalising details.'

... there are a few ex-military men who will back-up the writings of Mr. Allende, and thus corroborate the story, usually with the advantage of rather more detailed technical information. One, an Alfred D. Bielek, is these days a popular and entertaining speaker on the conference circuit, able to provide a wealth of information on his personal work on the Philadelphia Experiment (also known as Project Rainbow), including intriguing, if often baffling technical information, based on his considerable knowledge as an electronics engineer. Some cynics have pointed out that his recollection of the teleportation aspects only 'came back to him' after seeing the movie of the story, and that his photograph of a "Zero Time Reference Generator" looks strangely similar to an old Army field kitchen refrigeration unit. That said, he is certainly a man with a story to tell, and one which is too lengthy to cover here.

3. Official Navy documents. ... I do find the navy documents fairly convincing, but after the four different official explanations of the Roswell crash, I'm waiting to see if we get another official statement on the Philadelphia Experiment yet. What the navy documents do add, is some rather valid points on the subject of degaussing...

Anyway, this I find sounds suspiciously like the genuine seed for the story. Some sailor or other could quite accurately state, "They've got massive electro-magnets installed in the hull of the ship that make the ship invisible when we turn them on" or whatever, and hence set a rather elaborate rumour rolling, which really picked up steam whenever anyone could make money out of the story. The technical set up is remarkably similar to that described by Allende, though the effects are rather less outlandish! If I had to put my money on any version of the story, this would be the one I would go for, much as I would like to believe that time-space was bent that day.

...However, whilst faced with the eternal choice between believing an official military denial, and some cranky letters, many of us would choose the cranky letters 9 times out of 10. And let's face it—with good reason. The Military are professional deny-ers , and deny things as and when it suits them, even in the face of astonishing amounts of evidence. ...

All in all, the Philadelphia Experiment remains an enigma. It has all the traits of a classic piece of contemporary folklore—a root in historical fact, a suggestion of immensely exiting and dangerous scientific advances, possible military cover-ups, a hint of aliens (for good measure) and an almost total absence of reliable primary sources with which to ever prove what actually did happen. ...

A completely benign account was given by writer Jacques Vallee in an article in the *Journal of Scientific Exploration,* (Volume 8, Number 1, Spring, 1994) entitled "Anatomy of a Hoax: The Philadelphia Experiment 50 Years Later." He interviewed a sailor who was allegedly at sea with a sister ship of the *Eldridge* during 1943. The following can be found on the Internet at http:\\www.access.digex.net/~patin/philaj.html.

What Actually Happened in Philadelphia

In an earlier assessment of the Philadelphia Experiment data, the author offered the tentative conclusion that the story was, in part, based on fact: the Navy may have been involved in technically-advanced, classified tests in the Fall of 1943 (Vallee, 1991). These developments could have been misunderstood or deliberately romanticized by people like Allende, just as today we find tests of advanced flying platforms at Nellis Air Force Base being misinterpreted by believers. Furthermore I hypothesized that the experiments had to do with a radar countermeasures test. Indeed a Raytheon advertisement published thirteen years ago suggested that the corresponding technology was now out in the open (Raytheon, 1980). This hypothesis, however, failed to explain a few of the facts that highlighted the story. In particular it did not account for the observed disappearance of the destroyer from the harbor, for the mysterious devices brought on board under extreme security precautions, or for the alleged disappearance of two sailors from a nearby tavern. I called out to any one of my readers who might have additional information. That is how I came to correspond, and later meet face to face, with Mr. Edward Dudgeon.

"I am a sixty-seven year old retired executive. I was in the Navy from 1942 through 1945," began Mr. Dudgeon's letter (Dudgeon, 1992) explaining his purpose in contacting me.

He confirmed that the idea of an actual, secret technical development was correct, but he said I was wrong about a radar test. The truth, as he patiently wrote to me, was simpler.

"I was on a destroyer that was there at the same time as the *Eldridge* DE 173. ... I can explain all of the strange happenings as we had the same secret equipment on our ship. We were also with two other DEs and the *Eldridge* on shakedown in Bermuda and return to Philadelphia." ...

[An interview ensued]

"What was your training?" I asked him.

"I studied electronics at Iowa State. The Navy sent me to electronics school after boot camp. I graduated with the title of "electrician's mate third class" in February of 43, and then I went aboard ship in June 1943."

"Can you give me the name of the vessel?"

"Oh yes, the DE 50, *USS Engstrom*. It was a diesel electric ship, as opposed to the DE 173, the *Eldridge*, which was steam electric.

These ships were run by the electricians. Our ship was put in dry dock so they could install high-torque screws."

"Why the special equipment?"

"The new screws made a sound of a different pitch, which made it harder for the submarines to hear us. They also installed a new sonar for underwater detection, and a device we called a "hedgehog" which was mounted in front of the forward gun mount on the bow. It fired depth charges in banks of twenty-four to thirty in a pattern, and could cover 180 degrees as far as about a mile away.

That was one of the secrets. ... the Germans hadn't deployed radar at the time. We were trying to make our ships invisible to magnetic torpedoes, by de-Gaussing them. "In fact four ships were outfitted at the same time: the 48, 49, 50 and the *Eldridge*, in June and July of 1943. The Navy used to de-Gauss all the ships in dry dock, even the merchant ships, otherwise the vessels acted as bar magnets which attracted the magnetic torpedoes."...

"There was nothing unusual about the *Eldridge*. When we went ashore we met with her crew members in 1944, we had parties, there was never any mention of anything unusual. Allende made up the whole thing."

"What about the luminous phenomena he described?"

"Those are typical of electric storms, which are very spectacular. St. Elmo's fire is quite common at sea. I remember coming back from Bermuda with a convoy and all the ships being engulfed in what looked like green fire. When it started to rain the green fire would disappear."...

"Who was Allende? Did you ever meet him?" I asked, showing Mr. Dudgeon the various letters I had received from the man.

"I never did meet him. From his writings I don't think he was in the Navy. But he could well have been in Philadelphia at the time, serving in the merchant marine. He could also have been aboard a merchant ship we escorted back to the Philly-Norfolk area during a storm."

"What about the claim that generators were placed into the hold?"

"Aboard all diesel-electric and steam-electric destroyers there were two motors that turned a port or starboard screw. Each motor was run by a generator."

"What was the procedure when the Navy de-Gaussed a ship?"

"They sent the crew ashore and they wrapped the vessel in big cables, then they sent high voltages through these cables to scramble the ship's magnetic signature. This operation involved contract workers, and of course there were also merchant ships around, so civilian sailors could well have heard Navy personnel saying something like,"they're going to make us invisible," meaning undetectable by magnetic torpedoes, without actually saying it."

"What about the smell of ozone?"

"That's not unusual. When they were de-Gaussing you could smell the ozone that was created. You could smell it very strongly."...

"This doesn't tell us how the *Eldridge* disappeared into thin air, or what actually happened in the tavern in early August 1943."

"That's the simplest part of the whole story," Mr. Dudgeon replied.

"I was in that bar that evening, we had two or three beers, and I was one of the two sailors who are said to have disappeared mysteriously. The other fellow was named Dave. I don't remember his last name, but he served on the DE 49. The fight started when some of the sailors bragged about the secret equipment and were told to keep their mouths shut. Two of us were minors. I told you I cheated on my enlistment papers. The waitresses scooted us out the back door as soon as trouble began and later denied knowing anything about us. We were leaving at two in the morning. The *Eldridge* had already left at 11 p.m. Someone looking at the harbor that night have noticed that the *Eldridge* wasn't there any more and it did appear in Norfolk. It was back in Philadelphia harbor the next morning, which seems like an impossible feat: if you look at the map you'll see that merchant ships would have taken two days to make the trip. They would have required pilots to go around the submarine nets, the mines and so on at the harbor entrances to the Atlantic. But the Navy used a special inland channel, the Chesapeake-Delaware Canal, that bypassed all that. We made the trip in about six hours.

And so we have a convenient alternative explanation to the Philadelphia Experiment. Mr. Dudgeon's explanation of the degaussing procedures tie in well with the official Navy version. However, according to the Navy, the *Eldridge* did its training maneuvers in Bermuda in September and October of 1943, and did not go to Philadelphia thereafter. How did Mr. Dudgeon's ship accompany the *Eldridge* to Bermuda and Philadelphia in July and August? Mr. Dudgeon's account does not really explain the Allende/Allen letters, nor the mysterious deaths involved, including those of Morris Jessup and Ian Fleming. Is there more to the Philadelphia Experiment than the U.S. Navy and their paid researchers want us to know?

A Time Travel Reunion

As a final note on the history of the Philadelphia Experiment, it was reported in the *Bucks Courier Times* on Wednesday, March 24, 1999 that a time travel reunion of survivors of the Philadelphia Experiment would be having a reunion sometime that week.

Said the headline: "'Invisible' Ship Crew to Hold Reunion: About 15 members of a ship made famous by a supposed experiment in invisibility at the Philadelphia Navy Yard during World War II are gathering for the first time in 53 years."

No follow-up article has been reported, so we don't know whether the legend of the Philadelphia Experiment has been put to rest or not. Probably not. Myths and legends die hard, especially ones that span different periods of "time" as the Philadelphia Experiment and the

Montauk Project have been claimed to do.

Said one researcher into the whole mess, "It will all be exposed and uncovered—in due time."

Above: Physicist Albert Einstein in a meeting with Naval officers in his study in Princeton. Photo taken July 24, 1943. (National Archives)

Left: Photograph of the young Carl Allen on his Navy Certificate, dated March 4, 1944.

Photo # 306-PSG-51-686 (Box 45) Transferring DEs to Greece

Above: Official Navy photo, with caption, of the *USS Eldridge* upon its sale to the Greek Navy.

Top: The masthead of Gray Barker's column *Chasing the Flying Saucers* for his 1950s zine *Saucerian Bulletin,* the zine which first broke the story of Morris K. Jessup's death. Above: The only known photo of Morris K. Jessup, taken from an issue of Gray Barker's *Saucerian Bulletin.*

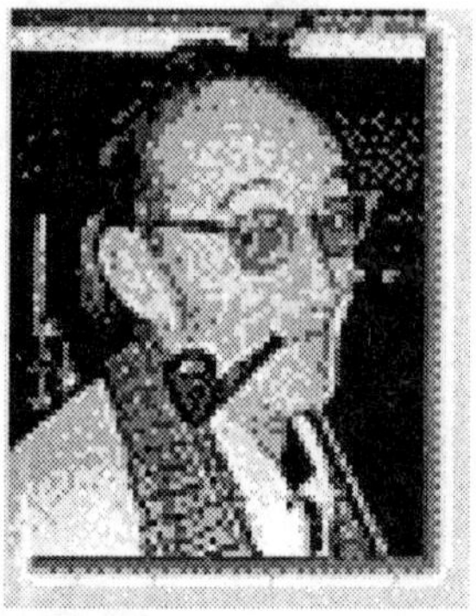

Left: An older Carl Allen at the NICAP offices in Tucson. Above: A photo of Allen/Allende in his later years when he lived in Colorado.

SERIAL NUMBER
A103664
BOOK NUMBER Z416175
UNITED STATES DEPARTMENT OF COMMERCE
BUREAU OF MARINE INSPECTION AND NAVIGATION
CERTIFICATE OF SERVICE TO ABLE SEAMAN

This is to certify that Carl Meredith Allen
18 years of age, born in Pennsylvania having given satisfactory evidence to the undersigned United States Board of Local Inspectors, Bureau of Marine Inspection and Navigation, for the district of NEW YORK that he has had the experience required by law, proof of which is made by affidavit, and having passed the examination as to eyesight, hearing and physical condition and the examination as to his efficiency, and having proved as a result of said examination that he has been trained in and is acquainted with the duties entitling him to the rating certification as prescribed by the Department of Commerce, is hereby rated as an able seaman for service on Any Waters

Issued by the undersigned Board of Local Inspectors on this 4th day of March 44
Raymond A. Shea
(Acting) Officer in Charge, Marine Inspection.
U.S. Board of Local Inspectors
Port of NEW YORK

Carl M. Allen
Signature of Seaman

Above: Navy Certificate, dated March 4, 1944, of Carl Allen with the number Z416175 that was used in the correspondence with Morris K. Jessup.

Above and below: Brilliant computer scientist Dr. John von Neumann (1903-1957), allegedly the brains, along with Einstein and Tesla, of the Philadelphia Experiment.

5.
The Philadelphia Experiment
The Technology

"Well, that's it, I guess I'm out of time."
—Al Bielek at the Global Sciences Conference
Daytona Beach, Florida, March, 1999

How can time travellers run out of time?
—Hatcher's Paradox No. IV

As noted in the last chapter, the public revelations of those purported to have taken part in the Philadelphia Experiment have drawn a lot of comment and further investigation. In this chapter, we will reprint some of this discussion relating to the actual technology claimed to have been used in the proceedings.

Al Bielek on the Zero Time Reference Generator and Time Locks

The following comments by Jerry Decker on Al Bielek's claims in his January 1990 speech were posted in the Mufon Metroplex article found at http:\\www.in-search-of.com:

"[One of] the two most outstanding things about the public lecture was the showing a slide purported to be a "Zero Time Reference Generator," which looks strangely similar to an old Army field kitchen refrigeration unit. No technical details are usually given.

"This device was purported to be the oscillator which drove the coils of the experiment. Mr. Bielek claimed that the unit shown is used to synchronize two separate signals (one for each coil). ...

"Basically, a coil was wound on each half of the ship and driven by separate oscillators, synced with an adjustable phase angle to create a "Scalar-type wave." This distorted the field matrices of matter encompassed within the field for "unusual effects." ...

"The other interesting comment was about a Professor of Mathematics at the University of Wisconsin in the 30's. His name—Henry Levenson. He specialized in time studies and developed a time variant equation.

"Mr. Levenson co-authored two books and wrote one. None of these books are available (of course) but Bielek indicates they might be available in restricted or private libraries such as Princeton.

"The time comments centered around a concept involving a TIME LOCK which was "encoded" at the time of creation of all matter, living or otherwise. Thus, all matter created on the Earth, must be "clocked" to the Earth time lock. The Earth must be "clocked" to the Solar time lock and that to the Galaxial.

"If your time lock became "distorted" by high intensity fields, it would create a variety of problems due to the instability of the recovery process, assuming that recovery could be possible.

"Bielek claimed that the original system was powered with a 500 KW generator, later increased to 2 MegaWatt. Another experiment was supposed to have been done with 3 field coils, all synced to the same clocking system. The 3-field design created major arcing and caused a return to the 2-field design. Bielek also claims that there were 3,000 tubes used in the system. ...

"The time theories espoused by Bielek seem particularly worthy of study. Much of the "Scalar" craze indicates a technology which could duplicate or surpass the original Philadelphia Experiment."

Another Teleportation Theory

The Mufon article continued with Jerry W. Decker adding his own theories on teleportation technology, basing it on UFO phenomena, the work of Dr. Harold S. Burr, Dr. Walter Russell and W. J. Hooper. Said the Mufon article:

> If the original purpose was "invisibility" as stated in various articles and commentaries, then why did a SHIP wind up in a HARBOR FILLED WITH WATER and NOT in the middle of a city street or a wheat field in Kansas.
>
> Seems that something is amiss as other accounts of the experiment state it was not carried out once but several times. Each time, the ship "teleported" to a watery target. What are the chances that EVERY TIME the ship would "land" in water??
>
> As I understand it, partially based on this quote from our file PHILAD1.ZIP: The U.S. Navy, for one, seemed obsessed with the

idea of the perfect camouflage—the ultimate secret weapon—INVISIBILITY. If only one of their warships could be made invisible, think of the havoc it could cause the enemy. Havoc that could conceivably bring the bitter and long-enduring war to an end.

On October 28, 1943, an experiment was conducted at the Philadelphia Navy Yard. This event, appropriately enough, became known as "The Philadelphia Experiment."

A Navy escort destroyer named the DE 173 (better known as the *U.S.S. Eldridge*), with hundreds of tons of electronic equipment aboard, lay in its dock. Scientists on shore started the experiment which involved Dr. Albert Einstein's "Unified Field Theory," a completed version appearing in the period 1925-27. "Withdrawn" as incomplete, the revised theory returned in 1940. The result of the experiment was brain-rattling!

The ship faded rapidly from sight in a foggy, green mist—and COMPLETELY VANISHED! Speechless amazement struck the wide-eyed scientists ashore.

Then, after a few minutes (another account makes that "seconds"), the ship reappeared where it had formerly been, to its Philadelphia dock, and regained its visibility! But that was not all. Something equally stunning was discovered to have happened during its vanishment. During that period of a few minutes (or seconds) the ship had shown up at its other Norfolk Navy Yard dock in Virginia! So, not only had it successfully become invisible, but it had also been TELEPORTED! Notice the arrows and how the ship went to its OTHER DOCK but in Virginia, not the one in Philadelphia.

Since we live in a serial time which clocks our every action and records it on the surrounding space continuum. Our presence at in any space at a given timeframe is captured by local space/matter conditions so that the moving of the earth and our solar system in its orbit would not leave us in space were we to teleport. So we have slices of wherever we have been since our creation. We should be able to step back into any given slice, not in time, only in space.

Another condition of this concept is that the PRESENT represents a suction which maintains matter as a balloon in the "holographic image" of a mass in a single time frame and in a specific spatial location. Remember that this is in the PRESENT ONLY.

Since the *Eldridge* had a history of having been at its OTHER DOCK SOMETIME IN THE PAST and probably on several occasions, then if we could distort the tempic (time) field enough, we could momentarily teleport the ship NOT THROUGH TIME BUT THROUGH SPACE to a place IT HAD BEEN BEFORE.

Now, if during that act and BEFORE COMPLETE TRANSFER WAS ACCOMPLISHED, someone cut the power, the portions of matter which had teleported would be "sucked" back like a vacuum to the source, i.e., the original spatial location of the PRESENT.

Think of it as two balloons. A full one in Philadelphia represents the *Eldridge*. At the port in Virginia, an empty balloon represents the holographic image (with no matter) and of a SPATIAL image sometime in the past.

(Of course, for each timeframe past, an empty holographic image balloon lies there recording the passage through time of each object, just like a frame of a comic strip.) If we slowly squeeze the full balloon (representing the *Eldridge* at the Philadelphia port in the original or PRESENT location), a connecting tube to the Virginia port allows the passage of "matter" into the empty *Eldridge*.

We get to a point where the source is 1/4 empty and the target is 3/4 full.

IF at this point, someone RELEASES THE PRESSURE, what happens? The "teleported matter" will surge back into the ORIGINAL SPATIAL location which exerts a vacuum to keep its balloon full.

Of course, once the transfer is complete, the suction then takes place from the target spatial location.

Since the pressure (high intensity pulsing magnetic waves) was released before the transfer was complete, the resulting surges created massive distortions in all mass encompassed by the field.

Living organisms record field effects in their tissues. Therefore, all tissue formations which took place during the surge, recorded the surges to "haunt" the victims. This apparently distorts the Bio-plasmic field also to cause major physical problems.

KeelyNet on the Zero Time Reference Generator and Nuclear Magnetic Resonance

Several good discussions of the Philadelphia Experiment technology were posted on the KeelyNet website (www.keelynet.com). The first we will reprint is the following written by Ken Anderson, an electronics technician:

Now for a couple of comments on Al Bielek. I wrote to him in the Fall of

1991, told him I'd been to Montauk Point, NY, and basically asked him to either tantalize me with some further specific technical details, or to point me toward some solid references on the subject. I figured, hey, if he was there and wants us to believe his incredible story, he'd be willing to answer a few technical questions. He never wrote back. I tried to track him down over the phone but was told by the Phoenix operator that his number is unlisted.

In July of 1991, Al Bielek gave a lecture at Rosemont College, PA, which I attended. He told his P.E./Phoenix story in substantially the same form as it was told on the now-famous videotape, "The Truth Behind the Philadelphia Experiment." What was new was that he had a BOOK, co-authored with Brad Steiger, called *The Philadelphia Experiment and Other UFO Conspiracies*. If you think the video causes brain damage, you should read the book! Anyway, as usual, it mainly told his story and other UFO anecdotes, but did not delve deeply into the PHYSICS or TECHNOLOGY end of the P.E. Bielek did say, however, that he was working on a new book that would TELL ALL from a technical standpoint, probably to be released in early 1992. Well, it's August 1992 at this writing. Has anyone seen or heard of this new book?

My research also unearthed some info about Bielek's "Zero-Time Reference Generator," which he claims Tesla invented and was used to set the "time locks" in P.E.-related experiments. Jerry Decker seems to imply that Bielek may have been hoodwinking his lecture audience when he showed a slide of this device, which Jerry remarks "looks strangely similar to an old Army field kitchen refrigeration unit. No technical details were given or offered."

Well, maybe I can offer some: A glance through my old Radio Shack 1975-76 *Dictionary of Electronics* shows me that there is indeed such a thing as a Zero Time Reference. Page 667 says, "In a radar, the time reference of the schedule of events during one cycle of operation."

In a digital or microprocessor circuit we would call it a "clock." Everything that goes on in the system is synced to that reference clock. Just what one might expect if they were indeed fooling around with TIME using radar-related equipment. Bielek made one other cryptic remark that bothered me for a long time—his remark on the E.T. Monitor program about a PI over 2 series of frequencies being WINDOWS, evidently INTERDIMENSIONAL windows. He never elaborated because it was the end of the program at that point. What was he referring to? Well, imagine my surprise when I ran across his terminology in the library one day while I was looking up info on Nuclear Magnetic Resonance! It seems that

"PI/2" refers to a certain pulse-width with respect to a reference pulse in one cycle of the NMR machine's field configuration. Also, Bielek refers to "T1" as being "our" normal time, while "T2" seems to be the "time" you're traveling to, in a time-travel experiment. Well, T1 and T2 also show up in the NMR literature, but apparently relating to the "relaxation time" between two events in the NMR cycle. Read the quotes (to follow) for details. So we have several possible scenarios here:

1) Al Bielek is a smart cookie who knows lots of technical terms gleaned from hi-technologies and appropriates them to his P.E. story, assuming that nobody is going to take the time to check out his lingo at the library,

2) He isn't necessarily borrowing terms from the NMR field, but, since the NMR pulse terminology probably derives from radar technology, he is using radar terminology to make his story sound impressive, or....

3) Perhaps Al is telling the truth after all. I just wish he would answer some real nitty-gritty technical questions from some qualified engineers so that the rest of us could know just how much of his story to believe, if any.

Without further adieu, let's look at what NMR is all about:

A compilation of info from several encyclopedias of Scientific terms, most notably *Van Nostrand's*.

"Nuclear magnetic resonance (NMR) is the effect of a resonant ROTATING (or alternating) magnetic field, imposed at RIGHT ANGLES to a typically much larger STATIC field, to perturb the orientation of nuclear moments. First applied to molecular beam studies, the term has come to refer specifically to the study of nuclear magnetism in bulk matter, with the distinct feature of detection by purely electromagnetic methods. The first such observations were made in 1946, for which F. Bloch and E.M. Purcell received the Nobel Prize in 1952...

"The essence of this methodology is that the nuclear moments u(i) in the specimen under observation are rotated so as to lie at some angle with respect to the applied field B(o). Under these conditions the moments experience a torque u(i) x B(o), which causes the angular momentum [i.e., u(i) itself] to precess around B(o) in precise analogy with the motion of a gyroscope in a gravitational field. The nuclei precess at the LARMOR PRECESSION FREQUENCY w(L)=gamma B(o), where gamma is the previously defined gyromagnetic ratio. For strong fields, w(L) is in the RADIO FREQUENCY RANGE. For example, in a field B(o)=1 T (=10^4 Gauss), protons precess at a frequency f(L)=w(L)/2PI= 42.5774 MHz. However established, coherently precessing nuclear moments can be caused to induce a voltage at frequency f(L) in a surrounding coil of wire, which can then be amplified by standard radio-frequency electronic techniques and recorded.

"Many techniques have been developed for the observation of NMR signals. These are divided into two principle classes, PULSED and CONTINUOUS WAVE (cw) methods. [Note that Al Bielek says Tesla favored the older CW method in the Philadelphia Experiment while John Von Neumann, his assistant, opted for the newer technology PULSED method.] In recent years the pulsed methods have gained wider acceptance because of their greater flexibility and efficiency. In pulsed NMR the signal following a single excitation pulse is known as a free induction decay. The excitation of free induction signals is accomplished by means of an applied ROTATING (or alternating) MAGNETIC FIELD of a frequency at or near the Larmor frequency of the nuclei to be studied, directed along an axis perpendicular to B(o)...

"Again, borrowing from the terminology of GYROSCOPIC MOTION, the rotation of nuclei about the field B(1) is generally referred to as NUTATION. If B(1) is applied long enough to rotate the nuclear moments into a RIGHT ANGLE with B(o) and then TURNED OFF, THIS IS REFERRED TO AS A PI/2 PULSE. A pulse twice as long, i.e., a PI pulse, will invert the nuclei to the direction -B(o). THE MAXIMUM FREE INDUCTION SIGNAL IS ACHIEVED WITH A PI/2 PULSE...

"Thus, nuclei in solids and liquids are typically very strongly decoupled from their surroundings, making them useful probes of local magnetic behavior. There are two principle types of relaxation process affecting NMR measurements. The first is longitudinal or spin-lattice relaxation, which is typified by the recovery of the nuclear polarization following the application of an excitation pulse. The time constant for the polarization to approach its equilibrium value is known as T1. In such a T1 process, the nuclear spins EXCHANGE ENERGY QUANTA... with the 'lattice' or environment in which they are immersed. [Bielek makes some references to 'cross-coupling of the lattices' in his story.] Depending on these circumstances, values of T1 range from the submicrosecond scale up to times of HOURS and beyond. Transverse relaxation is the term used to describe the decay in time of a free induction signal following pulsed excitation; it is characterized by a time constant T*2...

"The echo decay time constant is usually termed T2. ..."

—Van Nostrand's Scientific Encyclopedia
R.E. Walstedt

"Two-dimensional NMR. This method is a simple extension of the Fourier transform scheme to include a sequence of two PI/2 pulses separated by a variable time t1. Time t1, the 'evolution' time, is varied from zero up to values slightly greater than the free induction lifetime. After the second pulse, the signal is recorded over a time interval t2 with a similar range as t1. The data are then Fourier transformed over both time axes t1 and t2 to generate frequency scales f1 and f2, and a two-dimensional contour diagram of signal intensity against frequencies f1 and f2 is plotted.

"An application of NMR technique which has enormous potential for the fields of biology and medicine is that of NMR IMAGING OF SPECIMENS HAVING SPATIAL STRUCTURE.... The method has been developed to the point where cross-sectional images of the human body can be generated in just a few minutes with enough resolution to be of diagnostic value in medicine...

"A three-dimensional object to be imaged is placed in a magnetic field with a field GRADIENT along one axis, e.g., the z axis. A pulsed radio-frequency magnetic field is then applied, exciting the nuclear spins in a planar cross section of small but finite thickness oriented perpendicular to the z axis. This is the first pulse of the two-phase sequence described above. During time t1 a field gradient is switched on along the x axis, and after the second pulse, this gradient is switched to lie along the y axis. Barring excessive intrinsic structure in the spectrum of the species excited, the effect of these gradients will be to "encode" the spatial variation of NMR intensity into a spectrum of resonance frequencies, along one planar axis and then the other. On performing the Fourier transforms, this encoding will then result in a two-dimensional image of the spatially distributed NMR intensities."

As to the field strength of the main magnet, we read:

"MRI equipment consists of a huge, superconducting electromagnet that is cooled by liquid helium. It is constructed as a large tube with an inside diameter of about 1 meter to admit the patient. The electromagnet of a typical MRI system weighs more than 20 tons and is capable of producing a uniform magnetic field of 1.5 Tesla within its bore. (This is approximately 30,000 times the strength of the earth's natural magnetic field.)"

The preceding was a compilation of info from several encyclopedias of Scientific terms, most notably *Van Nostrand's*. One interesting side note: On the one hand we are not taught in school that there is any real danger in being exposed to strong magnetism (except for the current controversy over 60 Hz ELECTROmagnetism from power lines, etc.). Yet, elsewhere in this paper I included the quote about 'official' researchers 'frowning' on magnetic research because of possible deleterious effects to humans.

Well, here is a quote from an article on MRI that appeared in *Machine Design* magazine, 11/8/90:

"The magnet creates fringe field effects that unless taken into account by designers, stretch far beyond the MRI machine. Although there is no conclusive evidence of harmful effects of magnetic fields, the FDA, feeling it is better to err on the side of caution, has set up guidelines that no uninformed or unaware person should be permitted to walk through magnetic field strengths greater than 5 G... [The earth's field = about .6 Gauss]

"Large areas and parking lots surrounding early mobile units had to be roped off to prevent people from straying into the fringe field." —pp 76,77.

Just as food for thought: If you strayed too near a mobile MRI unit and soon afterward came down with some mysterious malady, do you really think you'd be able to prove that the magnetic field was the cause???

Well, there you have it. What are we able to glean from the descriptions of the technology behind the alleged Philadelphia Experiment? Was it a unique field/equipment configuration whose exact nature will remain hidden away in some Top Secret classified file in Washington? Or have the basic concepts, developed in the lab over the years under the name NMR, finally become popularly implemented since the late 1970's or early 1980's by the medical field, under the name MAGNETIC RESONANCE IMAGING? Or are the SCALAR theorists, such as Tom Bearden and Jerry Decker, correct when they imply that it's not the VECTOR electromagnetic fields that produced the Philadelphia Experiment effects, but the SCALAR component, produced by deliberately OPPOSING or CANCELING the "E" and "B" fields? I sincerely hope this rather long file will stimulate discussion, lots of responses, and maybe even some actual building of equipment along these lines. Last one to disappear into another dimension is a rotten egg!

The Philadelphia Experiment—Various Notes and Quotes

The following article by Rick Andersen was also posted on the KeelyNet web site, along with their editorial note:

Note: Rick Andersen is an electronics technician and experimenter based in Pennsylvania. This exhaustive essay on the "Philadelphia Experiment" ... is one of the most thoughtful and far-ranging pieces on the subject we've seen in a long time. Rick can be contacted at The Wrong Number BBS, 201-451-3063, 24 hrs, 19.2Kbps v.32bis/v.42bis/HST-or on the KeelyNet BBS, 214-324-3501, 24 hrs, 2400bps.

This file consists of a collection of quotations taken from several sources in an effort to piece together both the theory behind the Philadelphia Experiment and the method and equipment used to produce it. My interest in the Philadelphia Experiment is not only in its philosophical and technological impact on our world, should it happen to be a true story. As an electronics technician, I also have a keen interest in how they did it!

Here is an electronics—a physics—that they sure didn't teach us in tech school, or even in college! But the rumors persist, and I'm hereby challenging the writers out there, who keep dropping hints here and there, to do us all a favor and throw out a good, CONSISTENT, meaty bone for us to chew on.

People like Al Bielek make a big splash with an "I was there" announcement. The stories and anecdotes are absolutely spellbinding. The excursions into speculative Physics are sometimes so deep that the layman would not even know what questions to ask, much less be able to judge the truth or validity of the answers. And then you find out that he's still 'remembering' more details, and that his 'remembering' was given a good shot in the arm by his watching the 1984 movie, *The Philadelphia Experiment.*

Ahem.... So, if you're like me, and this thing sounds just plausible enough that you can't put it out of your mind, you begin to read and read everything you can get your hands on regarding the subject. Then you begin to read everything you can on indirectly-related, peripheral subjects, noting any possible connections to your main area of study. After a while you get a good feel for the similarities as well as the dissenting points of view as set forth by the various authors. And you wonder why everybody else "seems" to be informed about the how and the why, in great detail, but you. (I even traveled out to Montauk Point, Long Island, to explore the abandoned radar base which Bielek connects with both the P.E. and with "Project Phoenix." I plan to upload a file on that to KeelyNet soon.) You see, an electronics tech (and much more an Engineer, as Bielek is) knows damn well that there is a world of difference between DC and AC fields; between PULSED and ROTATING fields; between ELF and HF or MICROWAVE or RADAR frequencies; between VECTOR and SCALAR and STANDING WAVE versions of each. An electronics person knows that, without a DETAILED, comprehensive THEORY behind the bench set-up, he is not going to know how to set up voltages and currents, power levels, frequencies, waveforms, pulse widths or duty cycles. If there's a chance a circuit won't work, Murphy's Law dictates that it WON'T more often than not. Every electronics person reading this file knows what I'm talking about. In the KeelyNet file "Bielek-1.asc," Jerry Decker remarks that "The literature of the 'Philadelphia Experiment' is replete with descriptions of the technology."

Jerry offers a description: "Basically, a coil was wound on each half of the ship and driven by separate oscillators, synced with an adjustable phase angle to create a 'Scalar-type wave'. This distorted the field matrices of matter encompassed within the field for 'unusual effects'." Okay then, let's build one. What, no one has yet? This is the impression I got when I put in some telephone time searching for further info on the "Resonant Gravity Coil" whose construction plans can be found in the file "gravity3.asc." The author of the file calls himself "Shadow Hawk."

Anyone want to tell me how I can get in touch with Shadow Hawk? So I can ask him a question or two before I invest in 7500 feet of #16 magnet wire? (That's 1.42 MILES of wire!) What I'm saying is, let's not get so mesmerized by the theories and literary masterpieces and snake-oil salesmen floating around out there that our interest in "22nd Century" Science becomes solely an armchair pursuit, or, worse yet, a quasi-religion.

Now that my gripe has been unloaded, I'm ready to hazard a guess or two here in the hopes that somebody out there who knows something will be stimulated enough to write a file in response to this one, or adding to it. My speculation is this: There is a connection between the supposed technology of the Philadelphia Experiment and that modern medical imaging technique called NUCLEAR MAGNETIC RESONANCE.

The similarities between the descriptions of the two become more obvious when you go to the library and look up the subject of NMR (or MRI, Magnetic Resonance Imaging) in a science encyclopedia. The description of the technique and theory behind it are similar. The history of the development of NMR overlaps the P.E.—both are products of the 1940's. And, in certain respects the technology for both may well be an outgrowth of the development of RADAR, which also took place during the same period and which employs PULSED fields. In the pages that follow, I have assembled several quotes purporting to describe the Philadelphia Experiment—either with regard to its technology, or some anecdote which might shed some light on the story. Some of the quotes are not directly related to the P.E. but seem to be related in concept. ...

The occasional references to cyclones and cyclotrons tie in with rotating and/or vortex fields. Several quotes from one book are usually grouped together, separated from other groups by dashed lines. Although this file is a bit lengthy, I hope it serves to stimulate renewed attention to the exact technological methods used in the P.E. Even a small-scale duplication of the effect by an experimenter would tend to take the story out of the realm of myth, and establish its reality. Alfred Bielek claims to have been one of the two Navy seamen who threw the switches in the control room of the ship to power the equipment.

[*Time Travel Handbook* editor's note: We have rearranged a few of the following quotes included in Andersen's article, and inserted subtitles (in italics) regarding the sources of the quotes for easier reference by the readers of this book. We have deleted quotes of the Allende letters as they appear earlier in this book. All other editorial notes or segues in the following section are Andersen's or those of the authors quoted.]

Quotes from Al Bielek

Al Bielek gave the following description of the equipment to radio host Robert Barry in 1991:

"There were four RF transmitters and they were phased to produce a rotating field; they were pulsed at a 10% duty cycle. The magnetic component of the fields was generated by four large coils set on the deck of the ship, and they were run by two large generators down in the hold of the ship—75 KVAH—they were also pulsed. The entire system was under a very special control—a very peculiar type of control—that produced a rotating field effect, which produced the interaction, which caused the time field to shift." [frequency of pulsations unknown]

"[There are] FIVE DIMENSIONS [to our reality]: Our 3-D, time, and 'T2', which vector was being rotated in these experiments to give time/space shifts."

"All of the effects related to this are related to the PI over 2 series of frequencies, which are all WINDOWS." "Although those men above deck were made insane, if not 'fried', by the fields, out of this came an ELECTROMAGNETIC CANCER CURE that the Navy has 'sat' on for forty years and refuses to release to the world because that would be tantamount to admitting that the Philadelphia Experiment DID happen."
—Al Bielek, telephoning in to host Robert D. Barry on the television program "E.T. Monitor," June 1991. (WGCB, channel 49, Red Lion, PA.)

Quotes from Without A Trace by Charles Berlitz

"According to Jessup the purpose [of the Philadelphia Experiment] was to test out the effect of a strong magnetic field on a manned surface craft. This was to be accomplished by means of magnetic generators (degaussers). Both pulsating and non-pulsating generators were operated..." —Valentine as quoted by Berlitz in *Without A Trace*, p. 194.

"Jessup was worried about the experiments and told Valentine that the Navy had requested him to be a consultant on yet another experiment but that he had refused. He was convinced that the Navy, in seeking to create a magnetic cloud for camouflage purposes in October 1943, had uncovered a potential that could temporarily, and if strong enough perhaps permanently, rearrange the molecular structure of people and materials so that they would pass into another dimension with further implications of predictable and as yet uncontrolled teleportation." —*Without A Trace*, p. 201.

"I do not believe Dr. Jessup considered this an 'inadvertent' discovery. For many years, so I have been told, experiments involving high intensity magnetism have been officially discouraged, just as the ion motors, known for as far back at least as 1918, have been denied public disclosure and their inventors somehow silenced. I am therefore convinced that top-ranking physicists must have some knowledge—and understandable dread—of phenomena that might be expected to emerge from the generation of a high intensity magnetic field, especially a pulsating or vortextual one." QUESTION: "In the case of the alleged Philadelphia Experiment, is there a fairly simple scientific explanation as to what took place?" ANSWER: "To my knowledge there is no explanation in terms of the familiar or orthodox. Many scientists now share the opinion that basic atomic structure is essentially electric in nature rather than materially particulate. A vastly complicated interplay of energies is involved. Such a broad concept lends great flexibility to the universe. If multiple phases of matter within such a cosmos did NOT exist, it would be most surprising.

"The transition from one phase to another would be equivalent to the passage from one plane of existence to another—a sort of interdimensional metamorphosis. In other words, there could be 'worlds within worlds'. Magnetism has long been suspect as an involvement agent in such potentially drastic changes. To begin with, it happens to be the only inanimate phenomenon for which we have been unable to conceive a mechanistic analogue. We can visualize electrons traveling along a conductor and thus 'explain' electric current, or we can envisage energy waves of different frequencies in the ether and thus 'explain' the heat-light-radio spectrum. But a magnetic field defies a mechanical interpretation. There is something almost mystical about it. Furthermore, whenever we encounter incredible (to us) materialization and dematerialization, as in UFO phenomena, they seem to be accompanied by severe magnetic disturbances. It is, therefore, reasonable to suppose that a purposeful genesis of unusual magnetic conditions could effect a change of phase in matter, both physical and vital. If so, it would also distort the time element which is by no means an independent entity, but part-in-parcel of a particular matter-energy-time dimension such as the one we live in." —Valentine to Berlitz, *Without A Trace*, pp. 204, 205.

Quotes from The Philadelphia Experiment
by Charles Berlitz and William Moore

"My own special interest in the Philadelphia Experiment was connected with the possibility that a shift in the molecular composition of matter, induced by intensified and resonant magnetism, could cause an object to vanish-one possible explanation of some of the disappearances within the Bermuda Triangle."—Charles Berlitz, *The Philadelphia Experiment*, p. 14.

"The experiment [M.K. Jessup said to Dr. J. Manson Valentine] had been accomplished by using naval-type magnetic generators, known as degaussers, which were 'pulsed' at resonant frequencies so as to 'create a tremendous magnetic field on and around a docked vessel'." —*The Philadelphia Experiment*, p. 130.

"The experiment is very interesting but awfully dangerous. It is too hard on the people involved. This use of magnetic resonance is tantamount to temporary obliteration in our dimension but it tends to get out of control. Actually, it is equivalent to transference of matter into another level or dimension and could represent a dimensional breakthrough if it were possible to control it." —Jessup to Valentine, *The Philadelphia Experiment*, p. 131.

"In practice, it concerns electric and magnetic fields as follows: An electric field created in a coil induces a magnetic field at right angles to the first; each of these fields represents one plane of space. But since there are three planes of space, there must be a third field, perhaps a gravitational one. By hooking up electromagnetic generators so as to produce a magnetic pulse, it might be possible to produce this third field through the principle of resonance." —Valentine, *The Philadelphia Experiment*, p. 132.

"The thrust of [Einstein's attempt to produce a] Unified Field Theory was a string of sixteen incredibly complex quantities (represented by an advanced type of mathematical shorthand known as tensor equations), ten combinations of which represented gravitation and the remaining six electromagnetism...

"One thing that does emerge, interestingly, is the concept that a pure gravitational field can exist without an electromagnetic field, but a pure electromagnetic field cannot exist without an accompanying gravitational field." —*The Philadelphia Experiment,* p. 150.

"I think I heard they did some testing both along the river [the Delaware] and off the coast, especially with regard to the effects of a strong magnetic force field on radar detection apparatus. I can't tell you much else about it or about what the results were because I don't know. My guess, and I emphasize GUESS, would be that every kind of receiving equipment possible was put aboard other vessels and along the shoreline to check on what would happen on the 'other side' when both radio and low- and high-frequency radar were projected through the field. Undoubtedly observations would have also been made as to any effects that field might have had on light in the visual range. In any event, I do know that there was a great deal of work being done on total absorption as well as refraction, and this would certainly seem to tie in with such an experiment as this." —comments by anonymous military scientist, *The Philadelphia Experiment,* pp. 169,170.

"...The idea of producing the necessary electromagnetic field for experimental purposes by means of the principles of resonance was... initially suggested by [physicist] Kent... I recall some computations about this in relation to a model experiment [i.e., an experiment conducted using scale models rather than real ships] which was in view at the time... It also seems likely to me that 'foiling radar' was discussed at some later point in relation to this project...

"The initial idea seems to have been aimed at using strong electromagnetic fields to deflect incoming projectiles, especially torpedoes, away from a ship by means of creating an intense electromagnetic field around that ship. This was later extended to include a study of the idea of producing optical invisibility by means of a similar field in the air rather than in the water...

"He had on one part of a sheet a radiation-wave equation, and on the left side were a series of half-finished scratches. With these he pushed over a rather detailed report on naval degaussing equipment and poked fingers at it here and there while I marked with pencil where he pointed. Then Albrecht said could I see what would be needed to get a bending of light by, oh, I think 10 percent, and would I try to complete this enough to make a small table or two concerning it...

"I think that the conversation at this point had turned to the principles of resonance and how the intense fields which would be required for such an experiment might be achieved using this principle...

"Somehow, I managed to finish a couple of small tables and a few sentences of explanation and brought all back as a memo. We went in to Albrecht, who looked it all over and said, 'You did all this regarding intensities [of the field] at differing distances from the [ship's] beam, but you don't seem to pick up anything fore and aft'... All I had was the points of greatest curvature right off the ship's beam opposite this equipment...

"What Albrecht wanted to do was to find out enough to verify the strength of the field and the practical probability of bending light sufficiently to get the desired 'mirage' effect. God knows they had no idea what the final results would be. If they had, it would have ended there. But, of course, they didn't.

"I think the prime movers at this point were the NDRC and someone like Ladenburg or von Neumann who came up with ideas and had no hesitancy in talking about them before doing any computations at all. They talked with Einstein about this and Einstein considered it and took it far enough to figure out the order of magnitude he would need on intensity, and then spoke to von Neumann about what would be the best outfit to look into it as a practical possibility. That's how we got involved in it...

"I can also remember a point a little later when I suggested in a meeting of some sort that an easier way to make a ship vanish was a light air blanket, and I wondered why such a fairly complicated theoretical affair was under consideration. Albrecht took off his glasses at that point and commented that the trouble with having me at a conference was that I was good at getting them off the topic...

"I do remember being at at least one other conference where this matter was a topic on the agenda. During this one we were trying to bring out some of the more obvious—to us—side effects that would be created by such an experiment. Among these would be a 'boiling' of the water, ionization of the surrounding air, and even a 'Zeemanizing' of the atoms; all of which would tend to create extremely unsettled conditions. No one at this point had ever considered the possibility of interdimensional effects or mass displacement. Scientists generally thought of such things as belonging more to science fiction than to science in the 1940's. In any event, at some point during all this, I received a strong putdown from Albrecht, who broke in with something to the effect that 'Why don't you just leave these experimental people alone so they can go ahead with their project. That's what we have them for!'

"One of the problems involved was that the ionization created by the field tended to cause an uneven refraction of the light. The original concepts that were brought down to us before the conference were laid out very nicely and neatly, but both Albrecht and Gleason and I warned that according to our calculations the result would not be a steady mirage effect, but rather a 'moving back and forth' displacement caused by certain inherent tendencies of the AC field which would tend to create a confused area rather than a complete absence of color. 'Confused' may well have been an understatement, but it seemed appropriate at the time. Immediately out beyond this confused area ought to be a shimmering, and far outside ought to be a static field. At any rate, our warning on this, which ultimately went to NDRC, was that all this ought to be taken into account and the whole thing looked at with some care. We also felt that with proper effort some of these problems could be overcome... and that a resonant frequency could probably be found that would possibly control the visual apparent internal oscillation so that the shimmering would be at a much slower rate... I don't know how far those who were working on this aspect of the problem ever got with it...

"Another thing I recall strongly is that for a few weeks after the meeting in Albrecht's office we kept getting requests for tables having to do with resonant frequencies of light in optical ranges. These were frequently without explanation attached, but it seems likely that there was some connection here..." —comments by the anonymous elderly scientist to Moore in the chapter "The Unexpected Key," *The Philadelphia Experiment*, pp. 173-205.

Another testimonial, this time by an ex-military guard:

"I was a guard for classified audiovisual material, and in late 1945 I was in a position, while on duty in Washington, to see part of a film viewed by a lot of Navy brass, pertaining to an experiment done at sea. I remember only part of the film, as my security duties did not permit me to sit and look at it like the others. I didn't know what was going on in the film, since it was without commentary. I do remember that it concerned three ships. When they rolled the film, it showed two other ships feeding some sort of energy into the central ship. I thought it was sound waves, but I didn't know, since I, naturally, wasn't in on the briefing.

"After a time the central ship, a destroyer, disappeared slowly into a transparent fog until all that could be seen was an imprint of that ship in the water. Then, when the field, or whatever it was, was turned off, the ship reappeared slowly out of thin fog.

"Apparently that was the end of the film, and I overheard some of the men in the room discussing it. Some thought that the field had been left on too long and that that had caused the problems that some of the crew members were having." —reported to Moore by Patrick Macey, as told to him by a workmate named "Jim" in the summer of 1977, *The Philadelphia Experiment*, pp. 240, 241.

"The names of several scientists have come up in connection with the revival of such a project. Two government-employed scientists named Charlesworth and Carroll were reportedly responsible for installing the auxiliary equipment on the DE 173 and participated in the experiment, noting the neuronal damage 'due to diatheric' effect because of the magnetic oscillation of the magnetic field'." —*The Philadelphia Experiment*, p. 241.

"Victor Silverman, now living in Pennsylvania and still mindful of wartime security regulations and afraid of possible consequences, got in touch with the authors [Berlitz & Moore] through a third party when he first heard about the publication of a book about the DE 173. He speaks from personal experience: 'I was on that ship at the time of the experiment.'

"At the outbreak of WWII Silverman enlisted in the Navy. He, along with about 40 others, was destined to become part of a special secret Naval experimental project involving a destroyer escort vessel and a process which he could identify only as 'degaussing'. On board the vessel, Silverman noted that there was 'enough radar equipment on the ship to fill a battleship' including 'an extra mast' which was 'rigged out like a Christmas tree' with what appeared to be antenna-like structures.

"At one point during the preparation for the experiment, Silverman remembers seeing a civilian on board and said to a shipmate: 'That guy could use a haircut.' To his amazement he later discovered that the man had been Albert Einstein.

"Silverman was given the rating of Engineer, First Class, and, according to his account, was one of three seamen who knew where the switches were that started the operation. He also related that a special series of electrical cables had been laid from a nearby power house to the ship. When the order was given and the switches thrown, 'the resulting whine was almost unbearable.'..." —*The Philadelphia Experiment*, pp. 247, 248.

"This informant, who emphatically declined to be named, confided to Berlitz that he had seen highly classified documents in the Navy files in Washington, D.C., which indicated that at least some phases of the experiment are STILL in progress.

"In addition, scientific units in private universities, some possibly funded by the government, are reported to be pursuing research in magnetic teleportation, with the attendant invisibility as part of the experiment. Some recent reports place such experimentation as having taken place at Stanford University Research Facility at Menlo Park, Palo Alto, California, and at M.I.T. in Boston. However, in the words of one informant—M. Akers, a psychologist in San Jose, California—such magnetic experiments 'are frowned upon because they have detrimental effects on the researchers conducting the experiments.'"—*The Philadelphia Experiment*, p.255.

Quotes from Anti-Gravity and the Unified Field edited by David Hatcher Childress

John Walker, writing about Einstein's unified field concept in the book *Anti-Gravity & the Unified Field* says, "The U.S. Navy experimented with variations of the unifying vector on a large scale during WWII, in which they succeeded at least in part to make a ship invisible to radar. Radar is composed of ultra-high frequency waves, and here enters the problem. In order to make something invisible to ultra-high waves, you must take it beyond that, and now you are playing with an idea where time, space, and conventional geometry go right out the window, literally. The only other way is to create interference vectors BETWEEN yourself and the outside, a kind of shell as it were.

"The Navy, Einstein and a few others working together tried it the first way. They subjected the whole ship and crew inside a huge man-made field effect, then pumped its plane-rotational velocity, and raised the overall potential to beyond what might be considered the 'normal physics stability' threshold. Then came the book, followed by the movie, 'The Philadelphia Experiment'. Yes somebody out there is trying to tell us something, or make it so absurd that it falls into the entertainment category...

"In the large field effect experiment, generators were set up to produce a ROTATING MAGNETIC FIELD around the vessel while EXTRA MASTS were set up to accommodate the VERTICAL GRAVITY AXIS of the field. The had the equipment and some math to back up what was supposed to happen; however it's difficult to prepare human minds to get used to the idea of TRANSPOSING matter from one locale to another... The reported "freezing" of several of the crewmen...

"is what we should expect when inside a HYPER-FIELD, wherein all would become coherent or moving in the same direction. You would no longer be a 'man energy unit' aboard a 'vessel energy unit' upon the sea, but rather the electric fields of both man and vessel would occupy the SAME SPACE at SLIGHTLY DIFFERENT TIMES upon an almost perfect ground plane of water.

"There is also some evidence to support the [alleged] fact that in one of the experiments when ship and crew reappeared, some of the crew had merged bodily within the steel of the ship, just like we find when a tornado has passed. [See next reference to tornadoes/cyclones]. A rotary-motion field is the BASIS for particle-to-wave or wave-to-particle experimentation, for it is the TRANSITION POINT for energy to go either way...

"Matter as a product of energy [Einstein's famous equation E=mc squared]. Substance as a product of waves. Precipitate, gather the waves as in a funnel. The funnel causes rotary motion. Rotary motion causes a funnel. The motion creates a tight magnetic band similar to a torus around the apex of the funnel; the band itself is a spinning toroid which helps to keep what is gathered. All of this is unseen to the human eye. If precipitation continues on its course, the magnetic band becomes the horizontal equator. The vortex funnels become the vertical axis of the poles, of gravity. A particle is born and the principle is now a unit. Now we can see it.

"The spin has stabilized and slowed somewhat than at its birth. We can pump it up again, faster and faster; the particle will begin to radiate. If we keep this up the particle will 'de-centralize', will die as waves of heat, light and radio noise. Yes back to waves. This leads me to believe that the particle is simply a precipitation of wave coherence, that the wave is what is left after particle decay. They are separate only in their timeline of existence."

—John Walker, in *Anti-Gravity and the Unified Field*, pp. 59-62.

"There is a book entitled *Reality Revealed—The Theory of Multidimensional Reality* published in 1978 (Vector Associates)... [The authors, Douglas Vogt and Gary Sultan] refer to a 'cross talk' effect where time and space information from one locale are transmitted, via standing waves of a sort, to another locale where the overlap then 'unfreezes' into so-called normal time and space. Standing waves, they observe, can act as a high potential transmitter requiring very LOW POWER to initiate the action...

"The tornado passes and we are left with straw embedded in unshattered glass, a 2 x 4 piece of pine penetrating 5/8 inch steel, a 15 inch tire circling the base of a tree whose branches exceed 15 feet, and metal pipe UNDER the earth is left twisted in the wake of such funnels. Clearly something other than physical force AS WE KNOW IT manifests itself when CONDITIONS are met. I quote from *Reality Revealed*:

> "We contend that the tornado and the hurricane are examples of a cyclotron in reverse. [Here the reverse would be a high velocity (motion) creating a high frequency radio (wave) signal.]
>
> "At certain times of the year when the right temperatures exist, a giant capacitor is created. The earth is one plate and the upper atmosphere is the other plate. The earth's magnetic field envelopes these electrostatic plates. We theorize that when the earth is tilted at just the right angle, high-energy charged particles are actually able to enter the earth's magnetic field from space.
>
> "Because of the earth's tilt angle in reference to the direction of the high velocity particle, the particle is siphoned down to the tornado belts or hurricane areas. Here the right atmospheric conditions exist to form the electrostatic plates.
>
> "The high energy particle charges the plates of the capacitor and a damped, oscillating radio wave is created. High voltage standing waves are also created.
>
> "The damped, oscillating wave, along with the earth's magnetic field produces cyclotronic action of the atmosphere. In other words a nature-made cyclotron is created.
>
> "Because of the high electrical potentials created and the high standing waves produced, information that makes up the straw is translated (moved) in time; a translation in time is a translation in space. A translation in time of a microsecond of information at the speed of light represents a space translation of about 985 feet. Of course, if the straw just happens to be moved to the space occupied by a window or steel I-beam, it appears as if it has been blown

through these objects. But what really happened is that the straw and the I-beam occupied the same space, but at a different time, when the tornado was present. When the tornado passed by, the time for the straw and I-beam became one (unfreezes) again. The same thing can happen to living things."

Walker continues:

"The authors speak of the upper atmosphere and the earth acting as a giant capacitor in much the same way Tesla referred to it. The idea is not new but I contend that wherever there exist potential differences, there is a flow to balance that difference and this flow expresses its swirling nature. That this expression is the unifying factor between wave and particle physics. And that duplication and application of the unity factor leads to gravity control, teleportation, transmutation of elements, and time travel, to name a few."

—Walker, *Anti-Gravity and the Unified Field*, pp. 52-54.

Quoted References from a Chapter of the Book
Tapping the Zero-Point Energy by Moray B. King

Reference 43 to the chapter "Cohering the Zero-Point Energy," in the book *Tapping the Zero-Point Energy*, by Moray B. King, p. 100: "W.B. Smith, *The New Science*, Fern-Graphic Publ., Mississauga, Ontario, (1964). This esoteric work claims that the energized caduceus coil (opposing helical windings on a ferrite core) creates a 'tempic field'. [This is a region where the time component of the space-time metric is altered. Such a claim might be supported if the ZPE were a hyperspatial flux orthogonal to our three-dimensional space, and the pulsed magnetic field opposition on the ferrite lattice induced an orthorotation of this flux.]"

Reference 44, p. 100: "G. Burridge, 'The Smith Coil', *Psychic Observer*, 35(5), 410-16, (1979). This article explains how to wind a caduceus coil (previous reference) and reports on some observations made by investigators experimenting with the coil."

Reference 45, p. 100: "W.L. Moore, C. Berlitz, *The Philadelphia Experiment: Project Invisibility*, Grosset & Dunlap, NY, (1979). The authors research persistent rumors that in World War II the U.S. Navy did an experiment where, in attempting to bend light and radar waves around a ship by intense, pulsed magnetic fields, they accidentally teleported the ship. [If such an event did occur with technology of the 1940's, it would not have been overly complicated. Curving light around an object requires curving space-time. The only energy strong enough to do this would have to come from orthorotating the ZPE flux. Creating intense, pulsed, opposing magnetic fields with a caduceus coil is a candidate for accomplishing this. If too much ZPE were orthorotated, it would bend space-time too much, and the object would leave our three-dimensional continuum.]"

Reference 47, p.101: "S. Seike, *The Principles of Ultrarelativity*, G-research Laboratory, Tokyo, Japan, (1971). Seike proposes the existence of a physical hyperspace with an electrical energy flux that flows orthogonal to our three dimensional space. To orthorotate this flux into our space requires a four dimensional rotation. Seike calculates how the three dimensional projection of this 'hyper' rotation would appear in our 3-space using four dimensional Euclidian geometry. Seike's main theme is that by actualizing the dynamics of the 3-D projection with the motion of charge, the hyperspatial form is created."

Reference 48, p. 101: "C.W. Cho, *Tetrahedral Physics*, 449 Izumi, Komae City, Tokyo, Japan, (1971). Cho describes in detail one of Seike's (previous reference) hyperspatial, four-dimensional rotating forms called the 'resonating electromagnetic field' (RMF). The form is generated by rapidly switching electric charge in a specific way among four spheres located at the vertices of a tetrahedron. The dynamics of the switching are such that there are two modes of rotation orthogonal to each other: a rotational mode and a precessional, oscillating 'inside out' mode. The projected hyperspatial structure is described as a 'dynamical Klein bottle'. It is predicted that experimentally switching the charges in the described manner will cause the apparatus to exhibit gravitational and inertial anomalies."

Quotes from Ether Technology by Rho Sigma

Further info on Seike, from the book *Ether Technology: A Rational Approach to Gravity-Control,* by Rho Sigma (1977): [note the apparent incorrect spelling of his name: Seiki, rather than Seike, as in M.B. King's book.]

"Professor Shinichi Seiki, Uwajima City, Ehime Prefecture, Japan, developed a somewhat more elaborate theory of the Lorentz force, incorporating the use of the 'ether'. Starting with the so-called 'Kramer equation', which describes the movements of atoms in the presence of exterior electrical and magnetic fields—the basic components of the Lorentz force—Professor Seiki conceived of the possibility of creating 'negative gravitational energies' by utilizing a suitable electromagnetic field.

"Currently, in a process called NMR (Nuclear Magnetic Resonance), we only utilize the changes in spatial electron spins due to the application of magnetic fields. The substance to be examined is placed in a high frequency field, and we observe energy absorption effects peculiar to the frequencies typical of a given molecule of matter.

"Seiki went one step further and introduced NER (Nuclear Electrical Resonance), which influences BOTH the polar AND the axial spin. Polar spin, he claimed, is directly related to the gravitational field. Describing a rotating electrical AC field superimposed on a DC magnetic field, he claims that an exponential increase of 'negative gravitational energies' occurs at a certain resonance frequency. This means that energy from the earth-gravitational field enters the system of the secondary artificial field created by the antigravity motor. The negative G-energies cause a weakening of the earth-gravitational field, ultimately canceling it altogether. Further depolarization then causes the vehicle to be repulsed by the larger gravitational body (earth).

"It seems that the reason Prof. Seiki's NER effects have not yet been 'officially' utilized, is that nuclear electrical resonance can occur ONLY at extremely high electrical voltage SIMULTANEOUSLY WITH ultra-high AC frequencies. Below this threshold, the probability of negative-G-energy conditions is extremely small. Above this critical frequency (also called 'Larmor Frequency'), the effect of this type of gravity engine is also dependent upon the electromagnetic polarization potential of the materials used.

"Professor Seiki proposes ferromagnetic substances, such as ferrite and ferromagnetic materials such as barium-strontium-titanate. In his design, three spherical condensers are alternately charged and discharged by three magnetic coils. At first glance, the entire idea seems to be just another 'perpetuum mobile'. However, the only energy transformation used is that of gravitational energy into mechanical and electrical and vice versa."—pp. 82, 83. [Note that Tom Bearden's inventor friend, Floyd Sweet, also uses barium-ferromagetics in his "Vacuum Triode" free-energy device.]

End of Andersen article

An Alternate Explanation of the Philadelphia Experiment

On May 15, 1992, Rick Andersen posted the following alternative explanation on the KeelyNet bulletin board (www.keelynet.com). Andersen posted his explanation with free permission to use it in any publication without restriction. In the interest of time travellers everywhere, we reprint it in its entirety:

The "Philadelphia Experiment"—fact, legend, or whatever it is—continues to haunt those of us who have been hooked by that persistent tale of electromagnetic space-time warping that may have teleported a U.S. Navy ship in 1943. I, for one, want to crack this nut wide open, and so I'm always looking for new info on this and related subjects. In the classified ad section of *Popular Science*, October 1991, the following ad appeared: PHILADELPHIA EXPERIMENT: Ship rendered INVISIBLE (1943). Thorough TECHNICAL details U.S. $40. Alexander Strang Fraser, Box 991-C, Nelson, BC, Canada V1L6A5. Forever a fool for new tidbits, no matter how tiny, I sent Mr. Fraser the $40 he requested and waited for the technical secret of secrets. Two weeks later a small 4 x 6 inch package arrived.

I am reporting on it here for the benefit of all you good KeelyNetters who have better things to do with your $40 (like paying your phone bills, what with all that modem use!) than wasting it on a risk, when I am here to tell you what you would be getting—and all for FREE! (Well, minus the phone bill.) Why did I send $40 to Canada to someone I'd never heard of before? Because A. S. Fraser promised "thorough TECHNICAL details."

Holy smokes, maybe this guy was ON the *USS Eldridge*, or designed the electronic equipment, and has been hiding out in Canada all these years and is now ready to spill the beans for all of us hungry researchers!? Heck, I'll risk $40 to find out! My $40 investment turned out to be a small booklet entitled "Invisibility Technology" which, to quote Fraser, "includes concise specialized information that seems to be nonexistent in any other publicly accessible literature." Already I was breaking out in a cold sweat.

What could Fraser possibly have to reveal to us eager researchers? Fraser's booklet begins by rehashing the PE story as most of us have heard it, beginning with the "Allende" letters. He then presents a list of historical dates chronicling the Navy's wartime activities, as a background to the PE events. Fraser relates that the Navy was looking for a way to camouflage its ships (optically), which would serve to push the battleship back up to the forefront of our military arsenal; it seems that the development of aircraft and refinements of submarine technology were causing the battleship to lose some of its former importance and glory. Artificial camouflage would be a definite fix for that problem.

Alexander S. Fraser's main thesis is that the "field" created around the *USS Eldridge* (or whatever ship was used) was NOT electromagnetic in nature, contrary to what most of us have assumed; it was a THERMAL field. He asserts that any warping of space-time (a la Einstein) would have exhibited "enormous gravitational anomalies," which are absent from the usual PE accounts. Fraser interprets Einstein's theory to anticipate such gravitational 'shock waves' if space-time had actually been warped by massive electromagnetic fields.

He then concludes that, although INVISIBILITY is probably what the Navy was indeed trying to accomplish, the MAGNETIC WARP explanation is probably mere legend—perhaps first suggested by Carlos M. Allende's ASSUMPTIONS that the tests involved Einstein's Unified Field Theories. At this point Fraser looks at what Allende alleges to have WITNESSED, rather than simply speculated about Allende's remarks about a "scorch" field, fire, optical wavering like that seen through heated air-all components of Fraser's "thermal field."

Fraser speculates that the Navy was trying to induce an artificial OPTICAL MIRAGE effect via a blanket of heated air surrounding the ship. He compares the "optical wavering" effect to the mirage effects one can see across a hot pavement on a Summer day; similar atmospheric effects and "temperature inversions" have been known to cause certain islands to "vanish" optically when conditions are just right. How was this air blanket heated? Fraser cites scientific literature of the time to support his hypothesis that SONIC (sound) waves of high intensity were used to excite and heat the surrounding air molecules. Well, you certainly don't want the enemy to HEAR your sound source, so you must use an ULTRASONIC air horn—or siren, to be more exact.

Apparently, high power ultrasonic sirens were in use in the 1940's and it is interesting, as Fraser points out, that Carlos Allende's description of a pushing "flow" seems to fit the description of a "sonic wind" that would be experienced by anyone close to a high-power ultrasonic siren. Such a strong sonic field with its capacity to vibrate and heat matter (as in the ultrasonic cleaning machines used by industry today) would have detrimental effects on any crewmen nearby. Fraser even explains the mysterious "green haze" (similarly noted in Bermuda Triangle accounts) as being a result of exciting the surrounding sea water with powerful ultrasonics—"sonoluminescence" and related phenomena. But what about the reported TELEPORTATION of the *Eldridge* from Philadelphia to its other port at Norfolk, Virginia? Fraser speculates that some of the crew (on deck, only?) must have lost consciousness due to the powerful ultrasonic field—and so lost track of time. In other words, the ship DID go to Norfolk—only it SAILED there, it didn't TELEPORT!

To those crewmen who were out "cold," though, the trip seemed instantaneous. Fraser's brief "Appendix" does discuss more exotic possibilities such as gravity control, Zero-point Energy, etc., but he doesn't think that these were employed in the actual Philadelphia Experiment. As he suggests, the idea that Einstein's theories were the basis for the PE may "literally turn out to be a lot of hot air"! Fraser's ideas do seem to fit a number of aspects of the PE story. One gets the feeling that there are some good points to ponder, but also some rather flimsy assumptions—the "unconscious crew" theory being one that I can't buy too easily.

We should also consider the book by Berlitz & Moore, in which the anonymous elderly scientist (in the chapter, "The Unexpected Key") is quoted as saying that he, in fact, DID suggest to the Naval scientists that it would be easier to camouflage a ship by means of a "light air blanket," than by using high-power electromagnetics—but he was told curtly that the trouble with having him at meetings was that he was good at getting the others off the subject! And to 'let these scientific fellows get on with their work' or some similar statement. This anonymous elderly scientist had also voiced his concern over the "Zeemanizing" of the atoms of both ship and crew when the field was turned on.

The Zeeman Effect is defined as a spreading out of the spectral lines of atoms under the influence of a strong magnetic field. So this line of evidence appears to contradict Fraser's scenario. But Fraser's paper does present a balance, a counterpoint to some of the wild and unsubstantiated stuff being published about the Philadelphia Experiment. So from that standpoint I guess my $40 wasn't totally wasted. I just wanted to save my fellow KeelyNetters from blowing their hard-earned bread on a THEORY.

Alexander S. Fraser's ad implied that he had a TECHNOLOGICAL SCOOP. Instead, he delivers a SPECULATIVE explanation for the PE. His guess is as good as the next guy's. Mr. Fraser is $40 richer, I am $40 wiser, and you can contact me for copies of his report—for FREE—if you'd like to see his material firsthand. His booklet is copyrighted but I'm not selling it, so I'm not stealing his research for resale. My $40, however, is now in his pocket, which makes his report MINE, and if you request it from me, you'll at least know you're getting your money's worth!!

My address is: Rick Andersen, RD1 Box 50A, Newport, PA17074

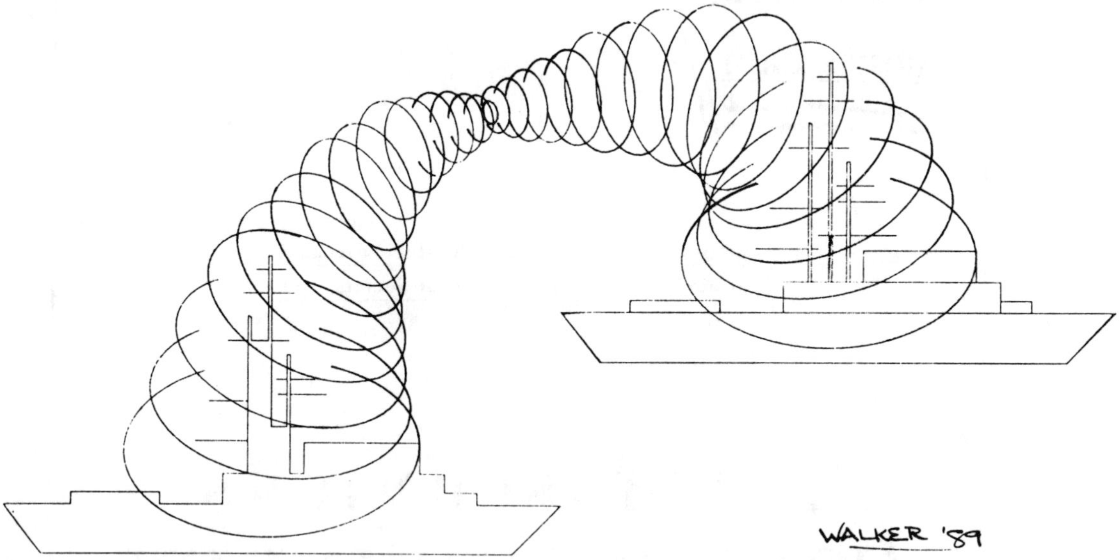

The whole ship and crew would have traversed space as a local electromagnetic tensor field wave dynamic.

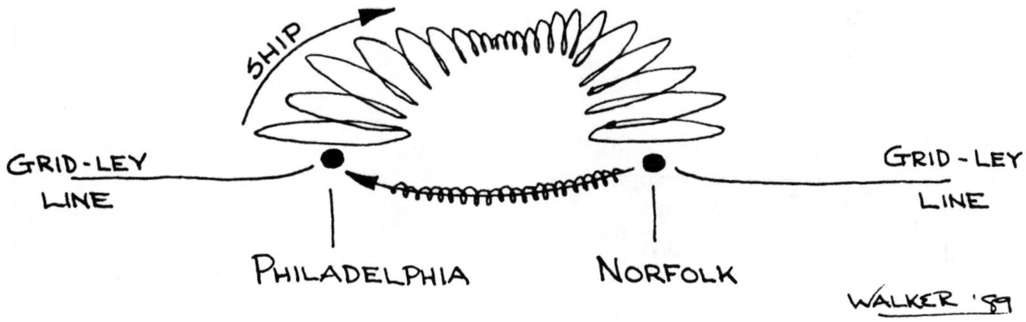

This illustrates how the ship, the DE 173, could have made its 200 mile jump as energy along a grid/ley line.

MOST SECRET

Copy No.

Attention is drawn to the Penalties attaching to any infraction of the Official Secrets Acts

INTER-SERVICES CODE-WORD INDEX

This Index will be kept in a safe when not in use

Issued under the authority of the Inter-Services Security Board, War Office.

1st *September*, 1941

C.C.S. 368/15.

When William Moore contacted the National Archives about Project Rainbow, they reported that no records existed of such a project or code name. However, when Moore obtained the Inter-Services Code-Word Index for 1941, it did indeed contain the code word "Rainbow," indicating that there may well have been an actual Project Rainbow as Carl Allen and Al Bielek have claimed.

Ciphering Numeral.	Code Word.	Block Number.	Ciphering Numeral.	Code Word.	Block Number.
7400	**PUSSYFOOT**	374	**7450**	**RACQUET**	266
7401	PUTNEY	590	7451	RADCLIFFE	748
7402	PUTRID	287	7452	RADFORD	854
7403	PUTTENHAM	804	7453	RADIAL	266
7404	PUZZLE	181	7454	RADIATION	324
7405	PYGMALION	164	7455	RADIATOR	491
7406	PYRAMID	170	7456	RADIO	2
7407	PYRITES	979	7457	RADISHES	304
7408	PYRRHUS	170	7458	RADIUM	430
7409	PYTHAGORAS	594	7459	RADNAGE	722
7410	PYTHON	288	**7460**	RADNOR	883
7411	QUADRANGLE	149	7461	RAEBURN	748
7412	QUADRUPED	548	7462	RAFFIA	266
7413	QUAGGA	619	7463	RAFFLE	422
7414	QUAGMIRE	126	7464	RAFTER	287
7415	QUAINTON	871	7465	RAGAMUFFIN	126
7416	QUARRY	143	7466	RAGMAN	181
7417	QUARTER	454	7467	RAGOUT	432
7418	QUARTERMAIN	388	7468	RAILWAY	33
7419	QUEEN	17	7469	RAIMENT	933
7420	QUEENBEE	643	**7470**	RAINBOW	334
7421	QUENCH	234	7471	RAINHAM	827
7422	QUICK	665	7472	RAISIN	229
7423	QUICKFIRE	104	7473	RAKEOFF	988
7424	QUICKLIME	548	7474	RAKISH	544
7425	QUICKSAND	561	7475	RALLY	188
7426	QUINCES	304	7476	RALPH	44
7427	QUINN	365	7477	RALSTON	827
7428	QUINTAL	123	7478	RAMBLER	431
7429	QUINTIN	266	7479	RAMESES	600
7430	QUIRK	585	**7480**	RAMIFICATION	63
7431	QUISLING	119	7481	RAMMER	288
7432	QUITMAN	659	7482	RAMONA	457
7433	QUIVER	129	7483	RAMOSE	158
7434	QUIXOTE	29	7484	RAMPAGE	434
7435	QUIZZY	173	7485	RAMPART	556
7436	QUOIT	393	7486	RAMPION	22
7437	QUORUM	563	7487	RAMPIRE	233
7438	QUOTATION	563	7488	RAMROD	143
7439	RABBITS	323	7489	RAMSDEN	853
7440	RABBLE	123	**7490**	RAMSHACKLE	561
7441	RABELAIS	450	7491	RAMSHORN	854
7442	RABID	652	7492	RAMSON	267
7443	RACEFIELD	559	7493	RANCHER	534
7444	RACHEL	447	7494	RANCID	141
7445	RACING	308	7495	RANDALSTOWN	738
7446	RACKETEER	487	7496	RANDOM	451
7447	RACKHAM	883	7497	RANELAGH	590
7448	RACONTEUR	266	7498	RANGE	19
7449	RACOON	266	7499	RANJI	601

The Pantagraph

Thursday, August 21, 1986 Bloomington-Normal, Illinois

Navy has lots of tricks to make carriers disappear

WASHINGTON, D.C. (AP) — U.S. Navy aircraft carriers, despite their incredible size, are becoming adept at a form of magic.

Utilizing weather, speed, advanced logistical planning and high-tech tomfoolery, several carriers in recent months have managed to vanish from antagonists' eyes into the vastness of the oceans, reappearing only at the moment of attack.

Last April, dogged by airplanes rented by American television networks and by Soviet intelligence vessels, the carriers Coral Sea and America dropped from sight off the coast of Sicily. Less than 24 hours later, their planes bombed targets in Libya.

And just over a month ago, a much lengthier case of a "missing" carrier occurred during an exercise named RIMPAC 86. The USS Ranger, although the target of an intense search that included satellite reconnaissance, escaped detection for two weeks while sailing across the Pacific.

The performance was considered all the more remarkable by an Australian admiral who monitored the exercise because the carrier's planes were flying sorties throughout the period, staging mock attacks against surface ships, submarines and land targets.

Rear Adm. I.W. Knox of the Royal Australian Navy disclosed recently the "Orange" forces in RIMPAC could not locate the Ranger "from the time it departed Southern Californian exercise areas until it steamed into Pearl Harbor some 14 days later."

Reports of such exploits delight Navy brass, who must answer critics who think carriers are sitting ducks in an age of nuclear-powered submarines and cruise missiles.

Modern-day carriers have yet to be tested in combat against Soviet weaponry. But they are practicing hard at what the Navy calls "maneuver strategy" — if the enemy can't find you, you have surprise. And with surprise, you can win.

Navy spokesmen decline to discuss the war-fighting tactics, citing military secrecy. But several officers interviewed recently, who asked not to be identified, say the idea of a "stealthy carrier" is not so far-fetched.

Consider:

- The Coral Sea and America accomplished their feats through a variety of tricks, but the most important were "masking" and "EMCON." The details of masking are classified, but essentially it involves making another ship — a destroyer, for example — look and sound like a carrier and a carrier look like something else.

Please see CARRIERS, page A5

CARRIERS

From A1

The process normally begins when a carrier is under radar surveillance, but beyond visual sight. The decoy ship maintains the carrier's previous course, while the carrier speeds away.

"We can make the Soviets believe another ship is the carrier," says one official. "The radar image, broadcasting pilot talk and the radio sounds of flight operations, the lighting at night: It looks like a duck and sounds like a duck so it must be a duck. So they follow the duck and make a mistake."

The carrier, meantime, can employ lighting at night that makes it look like a tanker.

• Also employed by the Coral Sea and America, and the key to the Ranger's disappearing act, was EMCON. This is the equivalent of a submarine "rigging for silence" or a convoy traveling under blackout conditions.

EMCON is a Navy acronym for emission control. Emission, in this case, refers to the electronic signals that are radiated by such equipment as radars, sonar and radio. When a carrier goes to EMCON, it literally shuts down much of its electronic gear to avoid detection.

Navy officials say a carrier can operate for long periods in EMCON because "we go mute, but not deaf or blind."

The procedure works by utilizing E2-C Hawkeye radar planes, flying at some distance from the carrier. Everything the Hawkeye sees is relayed electronically to the carrier and its escorts, providing a picture of aerial activity as well as surface forces.

While transmitting, the Hawkeye is far from the carrier, which gets the plane's signals passively without any transmission of its own. The Hawkeye also takes on the role of air-traffic controller for the carrier's planes.

Replenishment oilers, meantime, are told well in advance to make their own way to a specific position in the ocean. Again, radio silence is maintained.

• Aviation tactics. Even if radar can't pick up a carrier sailing beyond the horizon, the ship's location can be betrayed by jet aircraft scrambling into the air. The Navy's answer is called "offset vector."

"To be simplistic, the planes don't climb," says one officer. "They catapult off and literally hit the deck. If planes are suddenly popping up 100 miles from the ship, you have no idea where they came from."

— Speed. Publicly, the Navy says its carriers are capable of speeds "in excess of 30 knots." Privately, officers acknowledge the floating cities can approach 40 knots.

"We can literally outrun the Soviet tattletales (intelligence ships)," says one. "And in (heavy) weather of any kind, there's no contest. The carrier can outrun its own escorts."

— Weather and Satellites. Anyone who's been caught in the rain after the weatherman forecast sunny skies has his own thoughts on meteorology. But there have been solid gains made within that science in recent years.

"Although really heavy weather can hurt flight operations, these guys know how to follow weather patterns and use rain storms and above all, cloud cover," says one official. "The carriers can receive weather data via satellite, passively, without portraying their position."

"And we know the orbital parameters of Soviet reconnaissance satellites as well as our own," adds another. "If there's a recon bird coming by and you can duck into some weather, you duck into the weather. Or if you know there's a blind spot in coverage, you sail there."

"Once you succeed in slipping away," summarizes one officer, "the odds shift in your favor. Most people don't have any conception of how big the oceans are. You can be lonely if you want."

Above: Robert Butts stands next to a portrait of Seth.

6.

Time-Space Travel
According to Seth/Jane Roberts

by Madelon Rose Logue

"There are journeys of consciousness
that no one can take but you..."

"Space and time are intertwined."
—*Seth via Jane Roberts, 1977*

When I first read David Hatcher Childress' *Anti-Gravity and the World Grid,* I was so excited I let out a loud "Wow." The world grid maps included in the book showed the locations of the coordination points that Seth talked about in his book by Jane Roberts, *Seth Speaks.* He had written that the coordination points could be located mathematically, and here they were. Then, in 1989 I met David when I joined one of his Adventures Unlimited expeditions to look for a legendary lost city in the jungles of Peru. The brochure he sent me stated that each member of the expedition would "create his own adventure." That sounded so "Sethian" to me that I couldn't resist. When we didn't find the lost city I announced that it just hadn't yet been placed in our past from our future for us to find in the present. I said other stuff, too, like, our space scientists will sure be surprised when they find pyramids on the moon and Mars and discover that they are our own. I'd read about that in the Seth books by Jane Roberts way back in the early 70s. It was something I couldn't forget, I was so fascinated by the idea.

Jane Roberts started out writing poetry and science fiction. When she decided to write a book about ESP she and her husband, Robert Butts, borrowed a ouija board from their landlord and to their surprise, it worked. It wasn't long before Jane was in a trance state and channeling Seth while her husband, Rob, patiently sat and took down dictation. Seth has described himself as an "energy personality essence" no longer focused in physical form. He has also, and this one is my favorite because it gives me a good chuckle, described himself as a "bloodless old spook."

Together, they wrote over twenty books, including Jane's own two books of poetry, one children's book, and three science fiction novels (1) based not only on the Seth philosophy, but simultaneous time and something about ruins being placed in the past from the future to be found in the present. Jane Roberts became known world-wide as a modern day pioneer in the field of metaphysics and philosophy. Among the many subjects covered in her books are UFOs, ancient civilizations, and time-space travel.

When I received a letter from David telling me he was writing a new book about time travel and asking me if I would help him locate specific references in the Seth books, I thought, "Sure, no problem." Wow, was I wrong!

I spent two weeks searching. The early hardcover Seth books published by Prentice-Hall that I have are not indexed (although the newer editions published by Amber-Allen in San Rafael, California are), so I had only my own margin notes to go by. I asked fellow Seth readers if they recalled which book certain references were in. I went through my copy of *A Seth, Jane Roberts, and Robert Butts Combined Index,*(2) compiled by Bob Proctor. I found a lot of references to time, time-space, and space travel, UFO's, out-of-body travel, Atlantis, early civilizations, and probabilities, but not that one illusive sentence that stated that the pyramids on Mars and the moon were our own. However, I did locate the statement referring to the placement of things from the future into the past. It is in volume II of *The "Unknown" Reality,*(3) by Jane Roberts.

In Note 11 for Session 742, Seth says, "Rupert (Seth called Jane "Rupert" because he said it was her entity name) has implied in his novel, *The Education of Oversoul Seven,*(4), that some archaeological discoveries about the past are not discovered in your present because they do not exist yet. Now such concepts are difficult to explain in my kind of prose, and in your language. But in certain terms, the ruins of Atlantis have not been found because they have not been placed in your past yet, from the future.

"Now the future is probable. However, in your terms there are ruins of the civilizations that served as the 'concrete' basis for the one Atlantean legend. Those civilizations were scattered. The so-called ruins would not be found in any one place as expected, therefore. There are some beneath the Aegean Sea, and some beneath an offshoot of the Atlantic, and some beneath the Arctic, for the world had a different shape.

"...time is simultaneous, so those civilizations exist along with your own. Your methods of dating the age of the earth are very misleading.

"In your terms, from your present you 'plant' images, tales, legends, 'at any given time,' that seem to come from the past, but are actually like ghost images from the future, for you to follow or disregard as you choose."

I began wondering about myself. Had I only thought I'd read that bit about the pyramids on Mars being ours, or had I intuitively felt that would be the case someday? Or had I actually read the passage in a probable Seth book? That's not uncommon. Just about everyone I know who reads Seth books bumps into this one. Whenever we reread any of the books, it's like reading a brand new one because there's always something that wasn't noticed before and so it seems to be new material. We laugh about this all the time and tease each other with, "Oh, you must have read that in a probable book. I don't have that in mine." Here's an example: I'm sure I read just recently where Seth says that when we pick up a book to read, we create the words on the page as we go along. That's not an exact quote, and it's short and illusive, and probably not indexed, except perhaps in someone's own margin notes. It's the sort of statement that can stick in a person's mind like glue.

I could have simply assumed we'd find our own ruins on Mars after reading what Seth had to say about space travelers in chapter 15 of *Seth Speaks:*(5)

"...there have been other great scientific civilizations; some spoken of in legend, some completely unknown—all in your terms now vanished.

"...Groups of people in various cycles of reincarnational activity have met crises after crises, have come to your point of physical development and either gone beyond it, or destroyed their particular civilization.

"...they were given another chance... They began with a psychological head start as they formed new primitive groupings. Others, solving the problems, left your physical planet for other points in the physical universe. When they reached that level of development, however, they were spiritually and psychically mature, and were able to utilize energies of which you have no practical knowledge.

"Earth to them now is the legendary home. They formed new races and species that could no longer physically accommodate themselves to your atmospheric conditions.

"...They have discarded material form. This group of entities still takes a great interest in earth. They lend it support and energy. In a way, they could be thought of now as earth gods.

"On your planet they were involved in three particular civilizations long before the time of Atlantis; when, in fact your planet itself was in a somewhat different position.

"...The poles were reversed—as they were ... for three long periods of your planet's history. These civilizations were highly technological..."

Seth carries this further along in volume I of *The "Unknown" Reality*, (6) when discussing probable technologies in Session 702.

"...There have indeed been civilizations upon your planet that understood as well as you, and without any kind of technology, the workings of the planets, the positioning of the stars—people who even foresaw 'later' global changes. They used a mental physics. There were men before you who journeyed to the moon, and who brought back data quite as 'scientific' and pertinent. There were those who understood the 'origin' of your solar system far better than you. Some of these civilizations did not need spaceships. Instead, highly trained men combining the abilities of dream-art scientists and mental physicists cooperated in journeys not only through time but through space. There are ancient maps drawn from a 200-mile-or-more vantage point—these meticulously completed on return from such journeys.

"There were sketches of atoms and molecules, also drawn after trained men and women learned the art of identifying with such phenomena. There are significances hidden in the archives of many archaeological stores that are not recognized by you because you have not made the proper connections—and in some cases you have not advanced sufficiently to understand the information."

In the 40th session, 10:43, April 1, 1964, of *The Early Sessions,*(7) Seth said of space travel, "...In your terms it will take you too long to get where you want to go.

"Scientists will begin to look for easier methods ... the first really important discovery will be made by an orthodox scientist out of pure desperation. The scientific communities are even now being forced to consider the possibilities of telepathy as a means of communication ...

"It is very possible that you might end up in what you intend as a space venture only to discover that you have 'traveled' to another plane. But at first you will not know the difference."

Then in the 45th session, 9:35, (8) Seth continued, "Hypnotism will become more and more a tool of scientific investigation. Telepathy will be proven without a doubt, and utilized, sadly enough in the beginning, for purposes of war and intrigue. Nevertheless telepathy will enable your race to make its first contact with alien intelligence. It will not at first be recognized as such.

"(Here, Jane laughed as she paced back and forth in trance.)

"There is nothing more strange in such contact than there is in my contact with you. But because you are so involved with camouflage apparent reality, contact with such intelligence will be a startling discovery. The contact made will be from one male to another, although the alien male, from another camouflage galaxy, will be more involved than you consider possible.

"The actual telepathy contact with this alien intelligence will occur ... perhaps by the year 2001. However, for reasons that I will not go into, a hitch will develop of which your scientists will not be aware, at least in your terms. The intelligence that you contact will no longer inhabit that same universe by the time that the contact is made.

"By then you will have discovered that your present theory of the expanding universe is in error; and this error will ... affect your calculations as to the exact location in your space, of the intelligence contacted. The contact will be made, I believe, in Australia."

In 1997 a film, *Contact,* was made. Set in Australia, the movie is about a woman who makes contact with an alien.

"... Space travel will be dumped when your scientists discover that space as you know it is a distortion, and that journeying from one so-called galaxy to another is done through divesting the physical body of camouflage (matter). The vehicle of so-called space travel is mental and psychic mobility, in terms of psychic transformation of energy, enabling spontaneous and instantaneous mobility through the spacious present."

And then I came to the biggest mind-blower yet, and it's not in the later Seth books. Seth didn't like to make predictions because, as he often explains in his books, there really is no future, nor past, nor present, but simultaneous time, with everything happening all at once, and also, we do have free will. We create our own reality, a phrase originated with Seth and now in common usage. By using our inner senses, such as dreams, we have the ability to make changes, and to accept or reject any given predictions. This is why many psychic readings are wrong. The individual who receives the reading, changes the predicted outcome.

The Early Sessions is a set of six to ten books of dictation given by Seth prior to the first of the actual Seth books. As of this date, April 1999, four of these books are available from New Awareness Network, Inc., as it was Jane Roberts' last wish that all the work be published.

Beginning in November 1963, Seth was training Jane in the art of channeling and was giving all the background material for the series of books he would later dictate, with the intention of reminding the human race of what it had "forgotten" as he put it. He drew on all the previous work done by the great philosophers and metaphysicians and then pushed it all forward in an easy-to-understand language that would appeal to the more scientific western mind. He knew the world was ready. However, he kept Jane and Rob in the dark for quite a while as he led them "along the garden path" as Jane later noted. All of this was very difficult for her, and in reading *The Early Sessions* books, I've noticed how she rebelled. After all, she had wanted to become a good, mainstream writer of poetry and science fiction and not be involved in what she considered "New Age Crap." Reading the books is fun, informative, and mind-blowing. Try this one on from Session 55:(9)

"... Space travel, when it occurs, will utilize expansion of self. Your idea of death is based upon your dependence upon the outer senses. You will learn that it is possible, through no physical act, to relinquish the physical body, expand the self, using atoms and molecules as stepping stones to a given destination, and reforming the physical body at the other end."

Doesn't that sound a lot like "Beam me up, Scotty" from the old Star Trek TV series? Here's another little goody, "...any point in space is also a point in what you think of as time, a doorway that you have not learned to open." (10)

"True space travel would of course be time-space travel, in which you learned to use points in your own universe as 'dimensional clues' that would serve as entry points into other worlds.

"There are space-time coordinates that operate from your viewpoint—and space travel from the standpoint of your time, made along the axis of your space, will be a relatively sterile procedure... (Some reported instances of UFO's happened in the past as far as the visitors were concerned, but appeared as images or realities in your present. This involves craft sightings only.)," Seth stated in session 713 for October 21, 1974.(11)

In the previous session, 712 for October 16, 1974, Seth said, "...Your own coordinates close you off from recognizing that there are...other intelligences alive even within your own solar system. You will never meet them in your exterior reality...for you are not focused in the time period of their existence. You may physically visit the 'very same planet' on which they reside, but to you the planet will appear barren, or not to support life.

"In the same way, others can visit your planet with the same results."

So it seems the best way to travel and to meet others is to change our own focus to match each others' time periods. To do this we have to learn how to use our inner senses without fear and with as little distortion as possible. As Seth continues in the 712th session, "There are...inner coordinates having to do with the inner behavior of electrons. If you understood these, then such travel could be relatively instantaneous. The coordinates that link you with others who are more or less of your kind have to do with psychic and psychological intersections that result in a like space-time framework."

In January of 1964, in session 16, Seth expressed his surprise that we were able to see flying saucers at all and went on to explain something about their origins although he was not familiar with them in any depth.

"... beings from other planes (12) have appeared among you, sometimes on purpose and sometimes...by accident. As in some cases humans have quite accidentally blundered through the apparent curtain between your present and your past, so have beings blundered into the apparent division between one plane and another. Usually when they have done so they were invisible on your plane, as the few of you who fell into the past, or the apparent past, were invisible to the people of the past.

"...the flying saucer appearances come from...a plane...more advanced in technological sciences than earth at this time... the camouflage paraphernalia appears, more or less visible... The atoms and molecules that structurally compose the UFO, and which are themselves formed by vitality, are more or less aligned according to the pattern of its own territory. Now as the craft enters your plane a distortion occurs. Its actual structure is caught in a dilemma of form...between transforming itself completely into earth's particular camouflage pattern, and retaining its original pattern. The earthly viewer attempts to correlate what he sees with what he supposedly knows or imagines possible in the universe.

"What he sees is something between a horse and a dog, that resembles neither. The flying saucer retains what it can of its original structure and changes what it must. This accounts for many of the conflicting reports as to shape, size, and color. The few times the craft shoots off at right angles, it has managed to retain functions ordinary to it in its particular habitat.

"...These vehicles cannot stay on your plane for any length of time at all. The pressures that push against the saucer itself are tremendous...The struggle to be one thing or another is very great on any plane. To conform to the laws of a particular plane is a practical necessity, and at this time the flying saucer craft cannot afford to stay betwixt and between for any indefinite period...What they do is take quick glimpses of your plane..."

It is interesting to note that Seth was discussing UFOs in the 1960s and 1970s, and although he said beings from other planes and other planets had visited here, he did not say anything about actual physical contact taking place nor did he ever mention abduction as an actual physical experience. He always emphasized, in fact he repeated, over and over again, in every one of his books written through Jane Roberts, that we create our own reality and we get whatever we focus upon.

A friend of mine, who knew his metaphysics and understood the Seth material, once told me of a very vivid dream he'd had one night. In the dream he became aware of some aliens coming for him, and he knew they planned to abduct him and perform some sort of vile experiments on his person. They came so close he could feel them taking hold of him. Oh, he was scared. Then he remembered that he was dreaming, and he remembered that he created his own reality and that he created his own dreams. At that point in the dream, he declared his belief in his own power and willfully changed the reality of the dream event and then woke up.

This of course led to a long discussion from space-time travel into dreams as vehicles of travel and out-of-body experiences to reincarnation and simultaneous time.

Jane Roberts died in 1984 and though there are other people channeling Seth, they are channeling their own Seth, not Jane's Seth. There is a big difference.

Now we are into 1999—and much of what Seth told us about is coming to pass, or at least, is getting closer. Last year, at a gathering of Jane Roberts/Seth fans, Mill, who speaks through Harry Johnson in London, England, said (13) "...You will find within 50 or 100 years at most, but more likely 50, that vehicular transport will become no longer necessary, as you use your own ability to dissolve the pattern of your body and recreate it at a distance. Now, that appears ridiculous, something seen on television films at the moment. But this is a possibility, a probability: it will be a reality 70 years or so from now—in my estimation of probabilities... I am convinced that the most likely result of the passage of time will bring about the quality of mind which enables teleportation of the individual to occur. It is the only way by which other planets will be inhabited and other galaxies discovered and contacted.

"This is the future I shall project into. It is a future where people have time to cultivate their inner self as well as their outer self. It is a time of huge mental development, so that the children—and particularly the grandchildren—of people alive 20, 30 years from now will expect to be able to teleport themselves around the earth, to mentally communicate with each other.

"You do so at the moment... The inner self is very much involved in that sort of thing. And, as the inner self and the outer self merge, so a new quality of humanity will arise in bodies which are not separated into egocentric, selfish beings pretending to be divorced from the balance of their totality, but a race of whole people will be originated and will continue to spread and populate the earth. This is the kind of future to which I have committed myself and to which I sincerely hope you too will commit yourselves."

Mill, like Seth, is an "energy personality essence." However, Mill has never had an incarnation on our earth. He has been studying people through Harry Johnson, and plans to incarnate sometime in the next 25 - 30 years. Which means that we'll still be here and our earth will still be intact, no matter what the doomsday-sayers tell us. I think we are our own salvation, those of us who wish to be—and won't it be fun when we really can match our sense of time to that of another and go walk, not in the ruins of a legendary lost city in a jungle, but in a city still alive and not yet ruined and buried in the jungle?

BIBLIOGRAPHY

1. *The Education of Oversoul Seven,* 1973, *The Further Education of Oversoul Seven,* 1979, and *Oversoul Seven and the Museum of Time,* 1984, Prentice-Hall, Englewood Cliffs, NJ

2. *A Seth, Jane Roberts, and Robert Butts Combined Index,* compiled by Bob Proctor, copyright Robert F. Butts, 1996, *Seth Network International,* PO Box 1620, Eugene, Oregon 97440

3. *The "Unknown" Reality, volume Two, A Seth Book,* Jane Roberts, 1979, Prentice-Hall, Englewood Cliffs, NJ

4. *The Education of Oversoul Seven,* Jane Roberts, 1973, Prentice-Hall, Englewood Cliffs, NJ

5. *Seth Speaks, A Seth Book* by Jane Roberts, 1972, Prentice- Hall, Englewood Cliffs, NJ

6. *The "Unknown" Reality, volume One, A Seth Book,* Jane Roberts, 1977, Prentice-Hall, Englewood Cliffs, NJ

7. *The Early Sessions, Book 1,* Jane Roberts, 1997, New Awareness Network, Inc., PO Box 192, Manhasset, NY 11030

8. *The Early Sessions, Book 2,* Jane Roberts, 1997, New Awareness Network, Inc., PO Box 192, Manhasset, NY 11030

9. *The Early Sessions, Book 2,* Jane Roberts, 1997, New Awareness Network, Inc., PO Box 192, Manhasset, NY 11030

10. *The Nature of Personal Reality, A Seth Book,* Jane Roberts, 1974, Chapter 20, Session 671, 9:56 pm., Prentice-Hall, Englewood Cliffs, NJ

11. *The "Unknown" Reality, Volume Two, A Seth Book,* Jane Roberts, 1979, Prentice-Hall, Englewood Cliffs, NJ

12. *The Early Sessions, Book 1,* Jane Roberts, 1997, New Awareness Network, Inc., PO Box 192, Manhasset, NY 11030

13. "Create the Best Things in Life—They're Free," Barbara Waddell, *Black Sheep* #27, Feb/Mar 1999, edited and published by Madelon Rose Logue, Los Angeles, CA

14. *Anti-Gravity & the World Grid,* edited by David Hatcher Childress, 1987, Adventures Unlimited Press, Kempton, IL

Madelon can be reached at: Madelon Rose Logue 3868 Centinela Avenue #12 Los Angeles, CA 90066-4431 home 310-313-1162, Fax 310-452-5544, office 310-452-4953

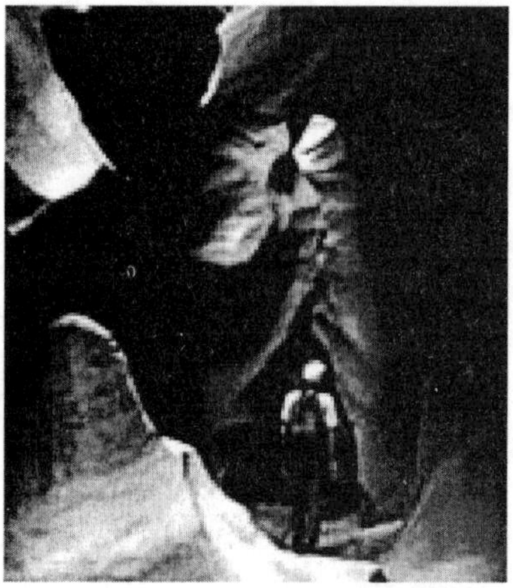

Images from the WingMakers web site: The canyon in Northern New Mexico, the petroglyphs, the entrance to the secret cave, and the inner room.

7.
THE WingMakers & TIME TRAVEL

The ACIO 1 Time Capsule

The Past is but the beginning of a beginning,
and all that is and has been is but the twilight of the dawn.
—*H.G. Wells, The Discovery of the Future (1901)*

"Truth is stranger than fiction
because fiction has to make sense."
—*This Is True (Internet Site)*

The following story was taken from the Internet. A web search for WingMakers will turn up the relevant pages that this story is based on. The editor cannot verify whether the following is true or not. What follows is the unedited transcript from "The WingMakers" posted in December of 1998:

Project Briefing and Background

In 1972, in a remote section of northern New Mexico, a group of hikers discovered an unusual artifact and pictographs within an obscure canyon. An archeologist from the University of New Mexico analyzed the artifact and searched the area where it was discovered, but found no signs that a prehistoric culture had established any permanent site in the canyon. It was presumed that a nomadic, Native American Indian tribe had occasionally used the canyon as a temporary settlement and had left behind a few artifacts of their presence as a consequence.

There were, however, two very puzzling questions. All but one of the artifacts could be dated to the 8th century AD. The exception, known as the "compass" artifact, appeared to be an unusual form of technology, and was found among more typical artifacts like pottery and simple tools. The compass was covered in strange hieroglyphic symbols, some of which were also found on the pottery. Secondly, the pictographs that were found in the area had inexplicably appeared, and they were strikingly different from any of the other native petroglyphs or rock art found in the southwest or the entire continent for that matter.

Because of these two anomalies, the artifacts and the entire project quickly became the property of the US government, or more specifically, the National Security Agency. It was decided that these artifacts might suggest a pre-historical, extraterrestrial presence on Earth, and that the NSA had the appropriate agenda and wherewithal to initiate a full-scale, scientific expedition to determine the nature and significance of the site.

The site was completely searched by a secret department of the NSA in 1973, but it only resulted in a few additional findings, and none of them were designated as technologies or evidence of an extraterrestrial presence. Additional pictographic symbols were found, but decoding them was a difficult and frustrating process. Experts were called in to help, but it was impossible to reach a consensus as to what the pictographs meant. As quickly as the project had risen as a priority investigation, it fell into the archives of the NSA under the code name, Ancient Arrow.

Twenty-one years later, in 1994, a series of rockslides opened up a section of the Ancient Arrow site. The canyon was in a naturally obscure section of park land held by the state of New Mexico. After its discovery in 1972, it had been officially sanctioned off-limits to hikers and campers and was to be left in its natural state. From time to time, scientists—sponsored by the NSA—would visit the site hoping to uncover new evidence, but were invariably disappointed.

Shortly after the rockslide occurrence, a small team of operatives from the NSA visited Ancient Arrow canyon to do some follow-up research. They discovered the rockslide had exposed an entrance to a hidden cavern that led deep within the canyon walls.

At the back of this cavern, the research team discovered a well-hidden entrance into the interior of the canyon wall or rock structure of the Ancient Arrow site. There they found a system of tunnels and chambers that had been carved out from solid rock. There were a total of 23 chambers, all intricately connected to an interior corridor, and each chamber held a specific wall painting, series of pictographs, written hieroglyphs, and what seemed to be dormant, alien technologies.

Once this entrance to the cavern was found, a report was immediately filed with the Director responsible for the Ancient Arrow project. The project was then formally brought under the jurisdiction of the Advanced Contact Intelligence Organization (ACIO), which organized an interdisciplinary research team to assess the exact nature of the site and attempt to discover additional artifacts or evidence of an extraterrestrial visitation.

The ACIO is a secret or unacknowledged department of the NSA. It is headquartered in Virginia, but also has personnel in Belgium, India, and Indonesia. They are largely unknown, even to senior directors within the NSA. The ACIO is the lowest profile organization within the entire intelligence community. Its agenda is to research, assimilate, and replicate any technologies or discoveries of extraterrestrial origin. Its personnel consist mainly of scientists who are completely anonymous, yet are paid salaries in excess of $400,000 per year because of their security clearance and IQ. This secret organization not only possesses enormous brain power, but it is also in possession of technologies that are far in advance of any other research facility on the planet. They are, in a word, privileged.

The artifacts found at the Ancient Arrow site were virtually incomprehensible to the research team. There were many mysteries. Why would an advanced culture leave their artifacts in such a precise and seemingly ordered manner? What was the message they were trying to leave behind? What were their technologies and why did they leave them behind? Did the creators of this site intermingle with the native tribes or remain an isolated culture? Who were they and why were they here in the 8th century? Were they planning to return? These were only some of the mysteries that challenged the research team.

Throughout the seven months of restoration, cataloging, and analysis, the Ancient Arrow project was a complete enigma. More energy went into the safe preservation of the artifacts, than attempting to solve the puzzle of their existence, though speculations were a topic of every conversation. Gradually, a hypothesis was structured by the research team that an extraterrestrial culture established an Earth colony in the 8th century and isolated itself within the Ancient Arrow canyon. They brought with them a very precise mission to leave behind a massive "time capsule" that would prove to be discovered in the late 20th century. While the exact nature of the time capsule was unclear to the research team, it seemed probable that it was a cultural exchange of some kind and had no invasive intent to Earth or its people.

It took a team of researchers nearly two years after the restoration was completed to decipher a partial meaning of the chamber artifacts. The 23 separate chambers seemed to be linked together to form some specific message or purposeful mission. In the 23rd and final chamber, they recovered a small optical disc that was presumed to hold digital information that could be the key to deciphering the artifacts. Scientists eagerly analyzed the disc, but they could not figure out how to access its content.

The ACIO's finest computer experts were called in to try and unlock the encoded disc, but to no avail. Several more months were spent trying every conceivable method to access the contents of the disc, but nothing worked. The Ancient Arrow project, for the first time in nearly a year, had hit a dead-end and funding for the project was rapidly weaned by the ACIO.

After two more months of unsuccessful efforts, it was decided that the technology to access the disc was simply not available. The optical disc and all of the artifacts and findings would be carefully placed in secure storage until the technologies were available to unlock the disc and harvest its content. It was presumed that the disc held star charts, translation indexes, glossaries, and all the answers to the various mysteries of its creators and, perhaps more importantly, their intentions for Earth.

While the optical disc was considered to be the key to unlocking the meaning of the time capsule, the ACIO had little choice but to place the project into storage and await the arrival of technologies that would permit them to unlock the disc. However, there were two scientists from the research team who theorized that the disc could be unlocked by understanding the meaning of the wall paintings in each of the 23 chambers. In their minds it was not a complex, technological solution, but rather a language or translation solution that would unlock the disc.

After much persuasion, the ACIO agreed to allow the two researchers to assemble a replica of the time capsule's contents. The replica time capsule consisted of detailed drawings and photographs of all the artifacts from each of the 23 chambers, including detailed, high-resolution photographs of the wall paintings. The two scientists would be allowed to continue their research on their own time provided they maintained utmost secrecy and reported all of their findings directly to the ACIO division head and project director.

The optical disc was securely stored away in a vault within the ACIO. The project was officially put on indefinite hold, and all personnel associated with the project were reassigned (with promotions) to different projects. The Ancient Arrow project was not to be spoken of again until which time the technologies—or some other means—provided a way to unlock the optical disc and access its contents.

The scientists spent nearly five months in partnership, trying unsuccessfully to decode the Ancient Arrow artifacts and establish the means to unlock the optical disc. During this time, the ACIO regularly experimented with new technologies or methods, and they too, were unsuccessful in unlocking the content of the optical disc.

One day, late in the summer of 1996, one of the scientists (a linguistics expert) had an insight into how to unlock the optical disc by reducing the symbols of the wall paintings to their closest facsimile found in an ancient Sumerian text. While the Sumerian language is extinct, it was sufficiently comprehensible to this scientist that he was able to decode the symbols of the paintings, and, placing the 23 words in the same order as the Ancient Arrow chambers, he was able to finally unlock the optical disc.

The connection between the Sumerian language and the time capsule was the breakthrough the ACIO team had been waiting for. A simple set of 23 words elicited over 8,000 pages of data from the optical disc. Unfortunately, the data was incomprehensible because there was no character set in the computer that could emulate the hieroglyphics and unusual symbols of the language. Thus, a translation index needed to be developed, which took an additional six months.

Finally, once a translation index was programmed into the computer, the data, while it could be printed out or viewed on the monitor in its hieroglyphic form, still required translation to English. And this translation process was extremely tedious and could not be facilitated through computers, owing to the subtlety of the language and its intricate connection to the wall paintings and pictographic representations elsewhere within the Ancient Arrow site.

As partial translations began to be developed, it was determined that even within the optical disc there was a segmentation of the data into 23 units. Each unit appeared to correspond to a specific chamber. As the first two chambers began to be translated, it was further shown that each unit contained philosophical and scientific papers, poetry, music, and an introduction to the culture and identity of its creators.

The creators of the time capsule referred to themselves as WingMakers. They represented a future version of humanity who lived some 750 years in our future. They claimed to be culture bearers, or ones that bring the seeds of art, science, and philosophy to humanity. They had left behind a total of seven time capsules in various parts of the world to be discovered according to a well-orchestrated plan. Their apparent goal was to help the next several generations of humans develop a global culture; a unified system of philosophy, science and art.

In early 1997, the ACIO scientist who had originally discovered the access code for the optical disc became strangely sympathetic with the WingMakers' mission. He was convinced that the ACIO would never share the discovery with the public, and he was certain that it was too significant to withhold. He also claimed that he was in communication with the WingMakers and that they were watching the ACIO's progress and would, at the appropriate time, make the time capsule and its contents available to the public.

This assertion troubled the ACIO and particularly the Ancient Arrow project director, who eventually recommended a leave of absence for the scientist who was summarily dispatched from the project. The scientist was afraid that his memories would be tampered with or destroyed altogether, and so he defected from the ACIO, quite literally the first to ever do so.

Soon after his defection, the scientist disappeared. However, before his disappearance, some of his materials regarding the WingMakers and their time capsule were given to a journalist that he had selected at random.

The author of this document is that journalist. I am in possession of photographs, music, poetry, artwork, translation indexes, copies of secret documents, and a variety of translated philosophical texts that all stem from the Ancient Arrow project. I've taken every precaution to remain anonymous so I can't be traced. I'm convinced that these materials are released against the wishes of a secret organization that probably has powers that even our government is unaware of.

Before the ACIO scientist (whom I will hereafter refer to as Dr. Anderson) had contacted me, I felt little or no interest in matters related to time travel, extraterrestrials, secret organizations, or anything else similar to these issues. When I initially heard the story it seemed preposterous, but I kept my journalistic objectivity, and met with Dr. Anderson and reluctantly concluded that it would be unlikely for an individual to fabricate this story with such detail and supporting evidence, and then desire to remain anonymous.

Dr. Anderson had brought files of photographs and drawings of odd-looking technologies that had strange symbols engraved on their outer casings. Research reports referencing the translation tables, cipher protocols, star charts, and dozens of memos from the ACIO department heads discussing the Ancient Arrow project. Everything, including about 400 pages of philosophical text had an authenticity to it that I was unable to reproach or ignore.

In fairness to those who will suggest I should investigate further in order to get independent corroboration before I present these materials, let me just say, that I'm unable to corroborate his story because of the very nature of the ACIO. However, for whatever reason, I trust Dr. Anderson who gave me these materials. He asked nothing from me. He desired no money or recognition. His only request was that I decide how best to bring these materials to the public. He counseled me not to investigate the ACIO because he was convinced the NSA would use misinformation tactics that would simply waste my time and make the goal of releasing these materials difficult if not impossible.

I've not contacted any other office of the government because Dr. Anderson told me that this would be traced by the ACIO who had high-level operatives in both the NSA and CIA, and, at best, would only invite misinformation tactics from one or both. I'm in possession of certain documents that I'll withhold from the WingMakers' web site, but if anything were to happen to me, I've arranged to have these documents shared with major media companies whom I know. These are my only safeguards in presenting these materials.

My only interest is in the release of these materials to the public, and then they can decide what to do about them. They may desire to pressure their politicians or take other action, it's their choice. I'm convinced that this story is too important to be held in the hands of an elite organization whose only interest is to re-engineer the technologies found in the Ancient Arrow site and apply them for their own agenda; no matter how noble that agenda may be.

I also realize that the Internet Service Provider who is hosting this web site may come under scrutiny, but if any pressure is exerted on this ISP to abandon its hosting service for this site, then this may also cause me to distribute the documents I referred to earlier. Let me be clear, these documents provide incontrovertible evidence of this secret organization known as the ACIO, and its elite directors are named and their real identities exposed.

I've spent the last several months agonizing about how these materials should be presented, and it seemed most appropriate to place them on the Internet to enable a global audience to access them. I have a close friend who created this web site whom I trust completely. Other than that, no one knows what I have done here (including my web site host).

You might ask why I've chosen to reserve full-scale media disclosure of the materials given to me by Dr. Anderson. I can only tell you that I don't want to create a circus atmosphere surrounding this discovery. It may ultimately end up in the mass media, but for now, my instincts are to keep a low profile for both these materials and myself. In doing so, I hope to preserve some sense of the dignity of these artifacts and let it grow from there.

I've never been involved in any story approaching this magnitude of importance, and I'm certain that if you spend some time on this web site and suspend your disbelief, even for a few minutes, you will see how important a discovery this time capsule is. The best way you can help is to spread the word about this discovery, and open the eyes of your political representatives. If you have web sites of your own, please link to the WingMakers' site.

Dr. Anderson had warned me that the ACIO has an advanced version of a technology based on what he called remote viewing. As I understand it, remote viewing is the ability to ascertain the whereabouts of people through some sort of "psychic insight" by someone trained in this technology. I know this sounds far-fetched, but Dr. Anderson was insistent that they had this capability and that it was one of their most feared technologies by those within the ACIO. In effect, it was known to keep their personnel loyal. Unfortunately, this will force me to stay underground and remain very mobile over the next several months, so don't expect too much change to the web site.

Believe me, I know that this whole story may seem impossible, but I can only tell you that I've seen detailed drawings and photographs of the artifacts taken from the Ancient Arrow site, and these are most assuredly, to my eyes, not of this time or world. They're unlike anything I've ever seen. Either the WingMakers are real, or someone has gone to a lot of trouble to convince me otherwise, and again, I'm a simple journalist without any ax to grind relative to secret government operations, ETs, time travel, or alien artifacts.

I'm not here to convert anyone. There is nothing to convert to. I simply want to disclose this material and let each individual absorb it as they choose. I will add additional documents and artifacts from the Ancient Arrow site when I feel it is safe to do so, but for now, there's enough material on this site to introduce anyone to the culture of the WingMakers.

I hope you take the time to immerse yourself in these materials. If you do, you may be surprised at the result.

Anne (not my real name)
Written October 23, 1998

The WingMakers and Time Travel Interview

In this section, you will find one of the five interviews conducted between Anne and Dr. Anderson back in December 1997, when Anne was first contacted about the discovery of the WingMakers' time capsule. These are the exact transcripts from her tape-recorded interviews and are probably the best way to understand the nature of the discovery and its implications.

What follows is a session I recorded of Dr. Anderson on December 27, 1997. He gave permission for me to record his answers to my questions. This was the first of five interviews that I was able to tape-record before he left or disappeared. I have preserved these transcripts precisely as they occurred. No editing was performed, and I've tried my best to include the exact words and grammar used by Dr. Anderson.

Anne: "Are you comfortable?"

Dr. Anderson: "Yes, yes, I'm fine and ready to begin when you are."

Anne: "You've made some remarkable claims with respect to the Ancient Arrow project. Can you please recount what your involvement in this project was and why you chose to leave it of your own freewill?"

Dr. Anderson: "I was selected to participate in the decoding and translation of the symbol pictures found at the site. I have a known expertise in languages and ancient texts. I am able to speak over 30 different languages fluently and another 12 or so languages that are officially extinct. Because of my skills in linguistics and my abilities to decode symbol pictures like petroglyphs or hieroglyphs, I was chosen for this task.

"I had been involved in the Ancient Arrow project from its very inception, when the ACIO took over the project from the NSA. I was initially involved in the site discovery and its restoration along with a team of 7 other scientists from the ACIO. We restored each of the 23 chambers of the WingMakers' time capsule and cataloged all of their attendant artifacts.

"As the restoration was completed, I became increasingly focused on decoding their peculiar language and designing the translation indexes to English. It was a particularly vexing process because an optical disc was found in the 23rd chamber and it was impregnable to our technologies. We assumed that the optical disc held most of the information that the WingMakers desired us to know about them. However, we couldn't figure out how to apply the symbol pictures found in their chamber paintings to unlock the disc.

"I decided to leave the project after I was successful in deducing the access code for the optical disc. Shortly thereafter I became aware of what I can only describe as the presence of the WingMakers. I felt as though they were visiting me... even assisting me in my work..."

Anne: "When you say 'visiting you', what evidence did you have that the WingMakers might be visiting you?"

Dr. Anderson: "I was spending 70 hours per week working on the decoding formulas for the symbol pictures, and this went on for about 8 months. During this time I tried every conceivable combination to create an access code to the optical disc. I was convinced it was the only way to open it. I was also convinced that it was purposely made to be difficult, at least to our present-day brains. It was almost as though the struggle to decode their language was exercising a part of my brain or nervous system that was enabling me to communicate with them.

"I began to hear them speaking to me. It began as a word or two... then a sentence... maybe just once a day. It didn't make much sense... what I heard. But then one day I was working on a chamber painting and I saw something move in the painting. One of the symbols moved and it was absolutely not an illusion or trick of the light. Then I realized that the WingMakers could interact with me, that they were time traveling to my time and that somehow their paintings were actually portals in which they moved through time.

"It was then I began to hear their instructions, or more precisely, their thoughts. I was given mental images on how to use the Sumerian language to decode their own symbol pictures. I thought I was possibly going crazy. I felt like my mind was playing tricks on me... that I was working too hard and needed to take a holiday, but I listened to the voices because it seemed plausible what I was being instructed to do. When I finished with the access code and it worked, I knew then that I was indeed communicating with them.

Anne: "Did you tell anyone? I mean about the fact that you were communicating with the WingMakers?"

Dr. Anderson: "I kept it a secret. I wasn't sure how I would be able to explain the phenomenon and I didn't want to arouse suspicions, so I went about my business and began developing the translation indexes for the 8,110 pages of text that were discovered within the optical disc. It was essential that we had a letter-for-letter index in order to retain the meaning of their language... we called this translation granularity. And as I started the process of translating the optical disc, I began to see fragment images of the WingMakers... sort of like a holographic image that would appear and then disappear in a matter of seconds.

"They visited me a total of three times—always in my home at night—and told me that I had been selected to be their liaison or spokesperson. Of course I asked them why me and not Fifteen, and they said that Fifteen was unable to speak for them because he was already the pawn of the Corteum."

Anne: "Tell me about Fifteen. What is he like?"

Dr. Anderson: "Fifteen is a genius of unparalleled intelligence and knowledge. He's the leader of the Labyrinth Group and has been since its inception in 1963. He was only 22 years old when he joined the ACIO in 1956. I think he was discovered early enough before he had a chance to establish a reputation in academic circles. He was a renegade genius who wanted to build computers that would be powerful enough to time travel. Can you imagine how a goal like that—in the mid-1950s—must have sounded to his professors?

"Needless to say, he was not taken seriously, and was essentially told to get in line with academic protocols and perform serious research. Fifteen came to the ACIO through an alliance it had with Bell Labs. Somehow Bell Labs heard about his genius and hired him, but he quickly out-paced their research agenda and wanted to apply his vision of time travel."

Anne: "Why was he so interested in time travel?"

Dr. Anderson: "No one is absolutely sure. And his reasons may have changed over time. The accepted purpose was to develop Blank Slate Technology or BST. BST is a form of time travel that enables the re-write of history at what are called intervention points. Intervention points are the causal energy centers that create a major event like the break-up of the Soviet Union or the NASA space program.

"BST is the most advanced technology and clearly anyone who is in possession of BST, can defend themselves against any aggressor. It is, as Fifteen was fond of saying, the freedom key. Remember that the ACIO was the primary interface with extraterrestrial technologies and how to adapt them into mainstream society as well as military applications. We were exposed to ETs and knew of their agenda. Some of these ETs scared the hell out of the ACIO."

Anne: "Why?"

Dr. Anderson: "There were agreements between our government—specifically the NSA—to cooperate with an ET species commonly called the Greys in exchange for their cooperation to stay hidden and conduct their biological experiments under the cloak of secrecy. There was also a bungled technology transfer program, but that's another story... However, not all the Greys were operating within a unified agenda. There were certain groups of Greys that looked upon humans in much the same way as we look upon laboratory animals.

"They're abducting humans and animals, and have been for the past 48 years... they're essentially conducting biological experiments to determine how their genetics can be made to be compatible with human and animal genetic structure. Their interests are not entirely understood, but if you accept their stated agenda, it's to perpetuate their species. Their species is nearing extinction and they're fearful that their biological system lacks the emotional development to harness their technological prowess in a responsible manner.

"Fifteen was approached by the Greys in his role at the ACIO, and they desired to provide a full-scale technology transfer program, but Fifteen turned them down. He had already established a TTP with the Corteum, and felt that the Greys were too fractured organizationally to make good on their promises. Furthermore, the Corteum technology was superior in most regards to the Greys... with the possible exception of the Greys' memory implant and their genetic hybridization technologies.

"However, Fifteen and the entire Labyrinth Group carefully considered an alliance with the Greys if for no other reason than to have direct communication with regard to their stated agenda. Fifteen liked to be in the know... so eventually we did establish an alliance, which consisted of a modest information exchange between us. We provided them with access to our information systems relative to genetic populations and their unique predisposition across a variety of criteria including mental, emotional, and physical behaviors; and they provided us with their genetic findings.

"The Greys, and most extraterrestrials for that matter, communicate with humans exclusively through a form of telepathy, which we called suggestive telepathy because to us it seemed that the Greys communicated in a such a way that they were trying to lead a conversation to a particular end. In other words, they always had an agenda, and we were never certain if we were a pawn of their agenda or we arrived at conclusions that were indeed our own.

"I think that's why Fifteen didn't trust the Greys. He felt they used communication to manipulate outcomes to their own best interest in favor of shared interests. And because of this lack of trust, Fifteen refused to form any alliance or TTP that was comprehensive or integral to our operations at either the ACIO or the Labyrinth Group."

Anne: "Did the Greys know of the existence of the Labyrinth Group?"

Dr. Anderson: "I don't believe so. They were generally convinced that humans were not clever enough to cloak their agendas. Our analysis was that the Greys had invasive technologies that gave them a false sense of security as to their enemy's weaknesses. And I'm not saying that we were enemies, but we never trusted them. And this they undoubtedly knew. They also knew that the ACIO had technologies and intellects that were superior to the mainstream human population, and they had a modicum of respect—perhaps even fear—of our abilities.

"However, we never showed them any of our pure-state technologies or engaged them in deep dialogues concerning cosmology or new physics. They were clearly interested in our information databases and this was their primary agenda with respect to the ACIO. Fifteen was the primary interface with the Greys because they sensed a comparable intellect in him. The Greys looked at Fifteen as the equivalent of our planet's CEO."

Anne: "How did Fifteen become the leader of both the ACIO and the Labyrinth Group?"

Dr. Anderson: "He was the Director of Research in 1958 when the Corteum first became known to the ACIO. In this position, he was the logical choice to assess their technology and determine its value to the ACIO. The Corteum instantly took a liking to him, and one of Fifteen's first decisions was to utilize the Corteum intelligence accelerator technologies on himself. After about three months of experimentation (most of which was not in his briefing reports to the then current Executive Director of the ACIO), Fifteen became infused with a massive vision of how to create BST.

"The Executive Director was frightened by the intensity of Fifteen's BST agenda and felt that it would divert too much of the ACIO's resources to a technology development program that was dubious. Fifteen was enough of a renegade that he enlisted the help of the Corteum to establish the Labyrinth Group. The Corteum were equally interested in BST for similar reasons as Fifteen. The Freedom Key, as it was sometimes called, was established as the prime agenda of the Labyrinth Group, and the Corteum and Fifteen were its initial members.

"Over the next several years, Fifteen selected the cream of the crop from the scientific core of the ACIO to undergo a similar intelligence accelerator program as he had, with the intention of developing a group of scientists that could—in cooperation with the Corteum—successfully invent BST. The ACIO, in the opinion of Fifteen, was too controlled by the NSA and he felt the NSA was too immature in its leadership to responsibly deploy the technologies that he knew would be developed as an outgrowth of the Labyrinth Group. So Fifteen essentially plotted to takeover the

ACIO and was assisted by his new recruits to do so.

"This happened a few years before I became affiliated with the ACIO as a student and intern. My stepfather was very sympathetic to Fifteen's agenda and was helpful in placing Fifteen as the Executive Director of the ACIO. There was a period of instability when this transition occurred, but after about a year, Fifteen was firmly in control of the agendas of both the ACIO and the Labyrinth Group.

"What I said earlier... that he was viewed as the CEO of the planet... that's essentially who he is. And of the ETs who are interacting with humankind, only the Corteum understand the role of Fifteen. He has a vision that is unique in that it is a blueprint for the creation of BST, and is closing in on the right technological and human elements that will make this possible."

Anne: "What makes BST such an imperative to Fifteen and the Labyrinth Group?"

Dr. Anderson: "The ACIO has access to many ancient texts that contain prophecies of the Earth. These have been accumulated over the past several hundred years through a network of secret organizations of which we are a part. These ancient texts are not known in academic institutions, the media, or mainstream society; they are quite powerful in their depictions of the 21st century. Fifteen was made aware of these texts early on when he became Director of Research for the ACIO, and this knowledge only fueled his desire to develop BST."

Anne: "What were these prophecies and who made them?"

Dr. Anderson: "The prophecies were made by a variety of people who are, for the most part, unknown or anonymous, so if I told you their names you would have no recognition. You see, time travel can be accomplished by the soul from an observational level... that is to say, that certain individuals can move in the realm of what we call vertical time and see future events with great clarity, but they are powerless to change them. There are also those individuals who have, in our opinion, come into contact with the WingMakers and are provided messages about the future, which they had recorded in symbol pictures or extinct languages like Sumerian, Mayan, and Chakobsan.

"The messages or prophecies that they made had several consistent strands or themes that were to occur in the early part of the 21st century, around the year 2011. Chief among these was the infiltration of the major governments of the world, including the United Nations, by an alien race. This alien race was a predator race with extremely sophisticated technologies that enabled them to integrate with the human species. That is to say, they could pose as humanoids, but they were truly a blend of human and android.

"This alien race was prophesied to establish a world government and rule as its executive power. It was to be the ultimate challenge to humankind's collective intelligence and survival. These texts are kept from the public because they are too fear-provoking and would likely result in apocalyptic reprisals and mass paranoia..."

Anne: "Are you saying what I think you're saying? That anonymous prophets from God knows where and when, have seen a vision of our future takeover by a race of robots? I mean you do realize how... how unbelievable that sounds?"

Dr. Anderson: "Yes... I know it sounds unbelievable... but there are diluted versions of this very same prophecy in our religious texts, it's just that the alien race is portrayed as the antichrist; as if the alien race was personified in the form of Lucifer. This form of the prophecy was acceptable to the gatekeepers of these texts, and so they allowed a form of the prophecy to be distributed, but the notion of an alien race was eliminated."

Anne: "Why? And who exactly is it who's censoring what we can read and can't? Are you suggesting there's a secret editorial committee that previews books before their distribution?"

Dr. Anderson: "This is a very complicated subject and I could spend a whole day just acquainting you with the general structure of this control of information. Most of the world's major libraries have collections of information that are not available to the general public. Only scholars are authorized to review these materials, and usually only on site. In the same way, there are manuscripts that were controversial and posited theories that were sharply different from the accepted belief systems of their day. These manuscripts or writings were banished by a variety of sources, including the Vatican, universities, governments, and various institutions.

"These writings are sought out by secret organizations that have a mission to collect and retain this information. These organizations are very powerful and well funded, and they can purchase these original manuscripts for a relatively small amount of money. Most of the writings are believed to be hocus-pocus anyway, so libraries are often very willing to part with them for an endowment or modest contribution. Also, most of these are original writings having never been published, being that they originated from a time before the printing press.

"There is a network of secret organizations that are loosely connected through the financial markets and their interests in worldly affairs. They are generally centers of power for the monetary systems within their respective countries, and are elitists of the first order. The ACIO is affiliated with this network only because it is rightly construed that the ACIO has the best technology in the world, and this technology can be deployed for financial gain through market manipulation.

"As for an editorial committee... no, this secret network of organizations doesn't review books before publication. Its holdings are exclusively in ancient manuscripts and religious texts. They have a very strong interest in prophecy because they believe in the concept of vertical time and they have a vested interest in knowing the macro-environmental changes that can affect the economy. You see for most of them, the only game on this planet that is worth playing is the acquisition of ever-increasing wealth and power through an orchestrated manipulation of the key variables that drive the economic engines of our world."

Anne: "So if they're so smart about the future, and they believe these prophecies, what are they doing to help protect us from these alien invaders?"

Dr. Anderson: "They help fund the ACIO. This collective of organizations has enormous wealth. More than most governments can comprehend. The ACIO provides them with the technology to manipulate money markets and rake in tens of billions of dollars every year. I don't even know the scope of their collective wealth. The ACIO also receives funding from the sale of its diluted technologies to these organizations for the sake of their own security and protection. We've devised the world's finest security systems, which are both undetectable and impregnable to outside forces like the CIA and the former KGB.

"The reason they fund the ACIO is that they believe Fifteen is the most brilliant man alive and they're aware of his general agenda to develop BST. They see this technology as the ultimate safeguard against the prophecy and their ability to retain relative control of the world and national economies. They also know Fifteen's strategic position with alien technologies and hope that between his genius, and the alien technologies that the ACIO is assimilating, that BST is possible to develop before the prophecy occurs."

Anne: "But why the sudden interest in the WingMakers' time capsule? How does it play a role in all of this BST stuff?"

Dr. Anderson: "Initially, we didn't know what the connection was between the Ancient Arrow project and the BST imperative. You have to understand that the time capsule was a collection of 23 chambers literally carved inside of a canyon wall in the middle of nowhere about 80 miles northeast of Chaco Canyon in New Mexico. It is, without a doubt, the most amazing archeological find of all time. If scientists were allowed to examine this site, with all of its artifacts intact, they would be in awe of this incredible find.

"Our preliminary assumptions were that this site was a time capsule of sorts left behind by an extraterrestrial race who had visited Earth in the 8th century. But we couldn't understand why the art was so clearly representative of Earth—if it were a time capsule. The only logical conclusion was that it represented a future version of humanity. But we weren't certain of this until we figured out how to access the optical disc and translate the first set of documents from the disc.

"Once we had a clear understanding of how the WingMakers wanted to be understood, we began to test their claims by analyzing their chamber paintings, poetry, music, philosophy, and artifacts. This analysis made us fairly certain that they were authentic, which meant that they were not only time travelers, but that they were also in possession of a form of BST..."

Anne: "Why did you assume they had BST?"

Dr. Anderson: "We believed it took them a minimum of two months to create their time capsule. This would have required them to open and hold open a window of time and physically operate within the selected time frame. This is a fundamental requirement of BST. Additionally, it is necessary to be able to select the intervention points with precision—both in terms of time and space. We believed they had this capability, and they had proven it with their time capsule.

"Furthermore, the technological artifacts they had left behind were evidence of a technology that was so far in advance of our own that we couldn't even understand them. None of the extraterrestrial races we were aware of had technologies so advanced that we could not probe them, assimilate them, and reverse-engineer them. The technologies left behind in the Ancient Arrow site were totally enigmatic and impervious to our probes. We considered them so advanced that they were quite literally indiscernible and unusable which—though it may sound odd—is a clear sign of an extremely advanced technology."

Anne: "So you decided that the WingMakers were in possession of BST, but how did you think you were going to acquire their knowledge?"

Dr. Anderson: "We didn't know, and to this day, the answer to that question is elusive. The ACIO placed its best resources on this project for more than four years. I posited the theory that the time capsule was an encoded communication device. I began to theorize that when one went through the effort to interact with the various symbol pictures and immerse themselves in the time capsule's art and philosophy, it affected the central nervous system in a way that it improved fluid intelligence.

"It was, in my opinion, the principal goal of the time capsule to boost fluid intelligence so that BST was not only able to be developed, but also utilized..."

Anne: "You lost me. What is the relationship between BST and fluid intelligence?"

Dr. Anderson: "BST is a specific form of time travel. Science fiction treats time travel as something that is relatively easy to design and develop, and relatively one-dimensional. Time travel is anything but one-dimensional. As advanced in technology as the Corteum and Greys are, they have yet to produce the equivalent of BST. They are able to time travel in its elemental form, but they can't interact with the time that they travel to. That is to say, they can go back in time, but once there, they cannot alter the events of that time because they are in a passive, observational mode.

"The Labyrinth Group has conducted seven time travel experiments over the past 30 years. One clear outcome from these tests is that the person performing the time travel is an integral variable to the technology used to time travel. In other words, the person and the technology need to be precisely matched. The Labyrinth Group, for all it knows, already possesses BST, but lacks the time traveler equivalent of an astronaut who can appropriately finesse the technology in real time and make the split-second adjustments that BST requires.

"The Labyrinth Group has never seriously considered the human element of BST and how it is integral to the technology itself. There were some of us who were involved in the translation indexes of the WingMakers, who began to feel that that was the nature of the time capsule: to enhance fluid intelligence and activate new sensory inputs that were critical to the BST experience."

Anne: "But I still don't understand what it was that led you to that conclusion?"

Dr. Anderson: "When we had translated the first 30 pages of text from the optical disc, we learned some interesting things about the WingMakers and their philosophy. Namely, that they claimed that the 3-dimensional 5-sensory domain that humans have adjusted to, is the reason we are only using a fractional portion of our intelligence. They claimed that the time capsule would be the bridge from the 3-dimensional 5-sensory domain to the multidimensional 7-sensory domain.

"In my opinion, they were saying that in order to apply BST, the traveler needed to operate from the multidimensional 7-sensory domain. Otherwise, BST was the proverbial camel through the eye of the needle… or in other words… impossible…"

Anne: "This at least seems plausible to me; why was it so hard to believe for the ACIO?"

Dr. Anderson: "This initiative was really conducted by the Labyrinth Group and not the ACIO, so I'm making that distinction just to be accurate, and not to be critical of your question. For Fifteen, it was hard to believe that a time capsule could activate or construct a bridge that would lead someone to become a traveler. This seemed like an extraordinarily remote possibility. He felt that the time capsule may hold the technology to enable BST, but he didn't believe it was merely an educational or developmental experience.

"The other outcome of immersion in the time capsule's contents was a sense of loyalty to the WingMakers' philosophy and approach to life. I found myself becoming less and less technology-centric and more and more spiritually focused. There was a sense of entrainment caused by their teaching that I couldn't explain. For whatever reason, I began to loose my objectivity as a researcher, and felt myself more of an advocate of the WingMakers."

Anne: "What do you mean by the word advocate?"

Dr. Anderson: "Just that I was sympathetic to what I construed as the WingMakers' agenda."

Anne: "And what was… or perhaps more appropriately, what is their agenda in your opinion?"

Dr. Anderson: "In my opinion, their agenda is to activate, through their time capsules the new consciousness that enables BST. I believe the WingMakers are trying to help us develop our consciousness… our human abilities… so we're able to utilize BST successfully as a defensive weapon. But more generally, I think this new consciousness is also—in itself—a defensive weapon."

Anne: "But if the WingMakers are time travelers themselves, in possession of BST, why can't they deal with the hostile aliens in 2011?"

Dr. Anderson: "I don't know. Believe me, I've thought about that one a great deal, as has the team working on the project. Perhaps BST isn't their primary concern for us, but rather helping us move from the 3-dimensional 5-sensory domain to the more potent multidimensional 7-sensory consciousness. Perhaps they're unable to access the intervention points because they lack some critical piece of information. Or perhaps they're unaware of the need because we already solved it in the year 2011.

"All I know is that we have about 6 different hypotheses, and we just don't have enough data to make a conclusion. Bear in mind that only about 7% of the text from the optical disc has been secured and translated to English. The ACIO is missing much of the information yet that will allow it to understand the true nature of the time capsules and the purpose of the WingMakers."

Anne: "Let's take a short break and resume after we've had a chance to grab some more coffee. Okay?"

Dr. Anderson: "Okay."

(Break for about 10 minutes... Resume interview)

Anne: "During the break I asked you about the network of secret organizations you mentioned that the ACIO is part of. Can you elaborate on this network and what its agenda is?"

Dr. Anderson: "There are many organizations that have noble exteriors and secret interiors. In other words, they may have external agendas that they promote to their employees, members, and the media, but there is also a secret and well-hidden agenda that only the inner core of the organization is aware of. The outer rings or protective membership as they're sometimes referred to, are simply window dressing to cover up the real agenda of the organization.

"The IMF, Foreign Relations Committee, NSA, KGB, CIA, World Bank, and the Federal Reserve are all examples of these organizational structures. Their inner core is knitted together to form an elitist, secret society, with its own culture, economy, and communication system. These are the powerful and wealthy who have joined forces in order to manipulate world political, economic, and social systems to facilitate their own agenda.

"The agenda, as I know it, is primarily concerned with control of the world economy and its vital resources—oil, gold, gas reserves, platinum, diamonds, etc. This secret network has utilized technology from the ACIO for the purpose of securing control of the world economy. They're well into the process of designing an integrated world economy based on a digital equivalent of paper currency. This infrastructure is in place, but it is taking more time than expected to implement because of the resistance of competitive forces who don't understand the exact nature of this secret network, but intuitively sense its existence.

"These competitive forces are generally businesses and politicians who are affiliated with the transition to a global, digital economy, but want to have some control of the infrastructure development, and because of their size and position in the marketplace can exert significant influence on this secret network.

"The only organization that I'm aware of that is entirely independent as to its agenda, and therefore the most powerful or alpha organization, is the Labyrinth Group. And they are in this position because of their pure-state technologies and the intellect of its members. All other organizations—whether part of this secret network of organizations or powerful multinational corporations—are not in control of the execution of their agenda. They are essentially locked in a competitive battle."

Anne: "But if this is all true, then is Fifteen essentially running this secret network?"

Dr. Anderson: "No. He's not interested in the agenda of this secret network. He's bored by it. He has no interest in power or money. He's only attracted to the mission of building BST to thwart hostile alien attacks that have been prophesied for 12,000 years. He believes that the only mission worth deploying the Labyrinth Group's considerable intellectual power is the development of the ultimate defensive weapon or Freedom Key. He's convinced that only the Labyrinth Group has a chance to do this before it's too late.

"You have to remember that the Labyrinth Group consists of 118 humans and approximately 200 Corteum. The intellectual ability of this group, aligned behind the focused mission of developing BST before the alien takeover, is truly a remarkable undertaking that makes the Manhattan Project look like a kindergarten social party in comparison. And perhaps I'm exaggerating a bit for effect... but I'm pointing out that Fifteen is leading an agenda that is far more critical than anything that has been undertaken in the history of humankind."

Anne: "So if Fifteen is running his own agenda, and it's just as you say it is, why would you defect from such an organization?"

Dr. Anderson: "The ACIO has a memory implant technology that can effectively eliminate select memories with surgical precision. For example, this technology could eliminate your recall of this interview without disrupting any other memories before or after. You would simply sense some missing time perhaps, but nothing more would be recalled... if that.

"My intuition cautioned me that I was a candidate to have this procedure because of the behaviors I was exhibiting in deference to the WingMakers. In other words, I was believed to be a sympathizer of their culture, philosophy, and mission—what I knew of it. That made me a potential risk to the project. The Labyrinth Group, in a very real sense, feared its own membership because of their enormous intellects and ability to be cunning and clever.

"This imprinted a constant state of paranoia which meant that technology was deployed to help ensure compliance to the agenda of Fifteen. Most of these technologies were invasive, and the members of the Labyrinth Group willingly submitted to the invasion in order to more effectively cope with the paranoia. Several months ago I began to systematically shut down these invasive technologies—in part to see what the reaction of Fifteen would be, and partly because I was tired of the paranoia.

"As I was doing this, it became obvious to me that the suspicions were escalating and it was simply a matter of time before they would ask me to subject myself to memory therapy. What I had learned from the WingMakers' time capsule is not something I want to forget. I don't want to give this information up. It has become a central part of what I believe and how I want to live out my life."

Anne: "Couldn't you have simply defected and not sought out a journalist who will want to get this story out? I mean, couldn't you have simply gone to an island and lived out your life and never disclosed the existence of the Labyrinth Group and the WingMakers?"

Dr. Anderson: "You don't understand... the Labyrinth Group is untouchable. They have no fears about what I divulge to the media; their only concern is the terrible precedence of defection. I'm the first. No one has ever left before. And their fear is that if I defect and get away successfully, others will too. And once that happens, the mission is compromised and BST may never happen.

"Fifteen and his Directors take their mission very seriously. They are fanatics of the first order, which is both good and bad. Good in the sense that they're focused and working hard to develop BST, bad in the sense that fanaticism breeds paranoia. My reasons for seeking out a journalist like you and sharing this knowledge is that I don't want the WingMakers' time capsules to be locked away from humanity. I think its contents should be shared. I think that was their purpose."

Anne: "This will seem like a strange question, but why would the WingMakers hide their time capsule and then encode its content in such an extraordinarily complex way if they wanted this to be shared with humanity? Look... if the average citizen had found this time capsule... or even a government laboratory, what's the chance they would have been able to decipher it and access the optical disc?"

Dr. Anderson: "It's not such a strange question actually. We asked it ourselves. It seemed clear to the Labyrinth Group that it had been the chosen organization to unlock the optical disc. To answer your question directly, had the time capsule been discovered by another organization, chances are excellent that its optical disc would never be accessed. Somehow, this coincidence—that the time capsule ended up in the hands of the Labyrinth Group—seems to be an orchestrated process. And even Fifteen agreed with that assessment.

Anne: "So Fifteen felt that the WingMakers had selected the Labyrinth Group to decide the fate of the time capsule's content?"

Dr. Anderson: "Yes."

Anne: "Then wouldn't it be reasonable to assume that Fifteen wanted to learn more about the contents of the time capsule before he released it to the public through the NSA or some other government agency?"

Dr. Anderson: "No. It's doubtful that Fifteen would ever release any information about the Ancient Arrow project to anyone outside of the ACIO. He's not one to share information that he feels is proprietary to the Labyrinth Group, particularly if it has anything to do with BST."

Anne: "So now that you've made these statements, isn't it going to affect the ACIO? Isn't someone going to ask questions and start poking around looking for answers?"

Dr. Anderson: "Perhaps. But I know too much about their security systems, and there's no way that a political inquiry will find them. And there's no way the secret network of organizations I mentioned earlier could exert any influence over them; they're completely indebted to the ACIO for technologies that permit them to manipulate economic markets. They... the ACIO and Labyrinth Group are, as I said before, untouchable. Their only concern will be defection... the loss of intellectual capital."

Anne: "What effect will your defection have on the ACIO or the Labyrinth Group?"

Dr. Anderson: "Very little. Most of my contributions with respect to the time capsule have been completed. There are some other projects having to do with encryption technologies that I developed and these will be more significant in their impact."

Anne: "Can you elaborate on the WingMakers and who you think they are or represent?"

Dr. Anderson: "I don't know who they are, but they represent themselves as human time travelers from the middle part of the 28th century. They could very well be the future version of the Labyrinth Group, or some other powerful organization. They seem to have a very well integrated sub-culture in that their language is clearly a combination of many extinct languages which they could only have knowledge of if they had access to ACIO information systems, or were indeed time travelers… or both, I suppose.

"Assuming they're accurately representing themselves, they are very advanced technologically. The Labyrinth Group holds that BST is the most advanced technology conceivable. Anyone who possesses it and can successfully utilize it, is clearly more advanced than our contemporary human culture or any of the extraterrestrials we are currently interfacing with."

Anne: "But if the WingMakers are so advanced technologically, why time capsules? Why not just appear one day and announce whatever it is they want to share? Why this game of hide and seek and hidden time capsules?"

Dr. Anderson: "Their motives are not clear. I think they planted these time capsules as their way to bring culture and technology from their time to ours. And they decided to do this by leaving behind these miraculous structures or time capsules that, once discovered, would lead people to a new philosophy or level of understanding. I think they're as interested in our philosophical outlook as our discovery of BST. Perhaps more so.

"As for why don't they just show up and give us the information… this, I think, is their genius. They've created seven time capsules and placed them in various parts of the world. I believe this is all part of a master plan or strategy to engage our intellects and spirits in a way that has never been done before. To demonstrate how art—culture, science, spirituality, how all of these things are connected. I believe they want us to discover this… not to be told.

"If they simply arrived here in your living room and announced they were the WingMakers from the 28th century, I suspect you'd be more amazed about their personalities and physical characteristics and what life is like in their time. That's assuming you even believed them. The aspects of what they wanted to impart—culture, art, technology, philosophy, spirituality, these items could get lost in the phenomenon of their presence.

"Also, in the text that I had translated, it was apparent that the WingMakers had time traveled on many occasions. They interacted with people from many different times and called themselves Culture Bearers. They were probably mistaken as angels or even Gods. For all we know, their reference in religious texts may indeed be frequent."

Anne: "So you think they intend that these time capsules be shared with the whole of humanity?"

Dr. Anderson: "You mean the WingMakers?"

Anne: "Yes."

Dr. Anderson: "I don't know with absolute certainty. But I think they should be shared. I don't have anything to personally gain from getting this information out to the public. It goes against everything I've been trained for and places me at risk and at the very least, disrupts my lifestyle irreparably.

"To me, the Ancient Arrow time capsule is the single greatest discovery in the history of humankind. Discoveries of this magnitude should be in the public domain. They shouldn't be selfishly secured and retained by the ACIO or any other organization."

Anne: "Then why are these discoveries and the whole situation with ETs kept from the public?"

Dr. Anderson: "The people who have access to this information like the sense of being unique and privileged. That's the psychology of secret organizations and why they flourish. Privileged information is the ambrosia of elitists. It gives them a sense of power, and the human ego loves to feed from the trough of power.

"They would never confess to this, but the drama of the ET contact and other mysterious or paranormal phenomena is extremely compelling and of vital interest to anyone who is of a curious nature. Particularly politicians and scientists. And by keeping these subjects in private rooms behind closed doors with all the secrecy surrounding it, it creates a sense of drama that is missing in most other pursuits.

"So you see, Anne, the drama of secrecy is very addictive. Now of course, the reason that they would tell you for keeping this out of the public domain is for purposes of national security, economic stability, and social order. And to some extent, I suppose there's truth to that. But it's not the real reason."

Anne: "Does our President know about the ET situation?"

Dr. Anderson: "Yes."

Anne: "What does he know?"

Dr. Anderson: "He knows about the Greys. He knows about ET bases that exist on planets within our solar system. He knows about the Martians…"

Anne: "Good God, you're not going to tell me that little green men from Mars actually exist are you?"

Dr. Anderson: "If I were to tell you what I know about the ET situation, I'm afraid I would lose my credibility in your eyes. Believe me, the reality of the ET situation is much more complex and dimensional than I have time tonight to report, and if I gave you a superficial rendering, I think you'd find it impossible to believe. So I'm going to tell you partial truths, and I'm going to be very careful in my choice of words.

"The Martians are a humanoid race fashioned from the same gene pool as we. They live in underground bases within Mars, and their numbers are small. Some have already immigrated to Earth, and with some superficial adjustments to their physical appearance, they could pass for a human in broad daylight.

"President Clinton is aware of these matters and has considered alternative ways to communicate with ETs. To date, a form of telepathy has been used as the primary communication interface. However, this is not a trusted form of communication, especially in the minds of our military personnel. Virtually every radio telescope on the globe has been, at one time or another, used to communicate with ETs. This has had mixed results, but there have been successes, and our President is aware of these."

Anne: "Then is Clinton involved in the secret network you mentioned earlier?"

Dr. Anderson: "Not knowingly. But he is clearly an important influencer, and is treated with great care by high-level operatives within the network."

Anne: "So you're saying he's manipulated?"

Dr. Anderson: "It depends on your definition of "manipulation." He can make any decision he desires; ultimately he has the power to make or influence all decisions relative to national security, economic stability, and social order. But he generally seeks inputs from his advisors. And high-level operatives from this secret network advise his advisors. The network, and its operatives, seldom gets too close to political power because it's in the media fish bowl, and they disdain the scrutiny of the media and the public in general.

"Clinton, therefore, is not manipulated, but simply advised. The information he receives is sometimes doctored to lead his decisions in the direction that the network feels is most beneficial to all of its members. To the extent that information is doctored, then I think you could say that the President is manipulated. He has precious little time to perform fact checking and fully evaluate alternative plans, which is why the advisors are so important and influential."

Anne: "Okay, so he's manipulated—at least by my definition. Is this also happening with other governments like Japan and Great Britain for instance?"

Dr. Anderson: "Yes. This network is not just national or even global. It extends to other races and species. So its influence is quite broad, as are the influences that impinge upon it. It is a two-way street. As I said before, the Labyrinth Group operates the only agenda that is truly independent, and because of its goal, it's permitted to have this independence... though in all honesty, there's nothing that anyone could do to prevent it, with the possible exception of the WingMakers."

Anne: "So all the world's governments are being manipulated by this secret network of organizations... who are these organizations... you mentioned some of them, but who are the rest? Is the mob involved?"

Dr. Anderson: "I could name most of them, but to what end? Most you wouldn't recognize or find any reference to. They are like the Labyrinth Group. Had you ever heard of it before? Of course not. Even the current management of the NSA is not aware of the ACIO. At one time, they were. But that was over 35 years ago, and people circulate out of the organization, but still retain their alliance to the secret and privileged information network.

"And no, absolutely there is no mob or organized crime influence in this network. The network uses organized crime as a shield in some instances, but organized crime operates through intimidation, not stealth. Its leaders possess average intelligence and associate with information systems that are obsolete and therefore non-strategic. The organized crime network is a much less sophisticated version of the network I was referring to."

Anne: "Okay, let's get back to the WingMakers for a moment... and I apologize for my scattered questions tonight. It's just that there's so much I want to know that I'm finding it very difficult to stay on the subject of the Ancient Arrow project."

Dr. Anderson: "You don't need to apologize. I understand how this must sound to you. I'm still wide awake, so you don't have to worry about the time."

Anne: "Okay. Let's talk a little bit about your impressions or insights into the WingMakers' philosophy and culture."

Dr. Anderson: "First of all, again, I want to remind you that only a fraction of their writings have been translated. So whatever insights I may have, are limited by a partial understanding—at best—of their culture and philosophy. Also, I want to remind you that the WingMakers may not represent the broader culture and philosophy of their time. Our interpretation was that they represented a subset or subculture of their time.

"With those qualifications, I'll say that the WingMakers have the benefit of about 750 additional years of evolutionary thought. We presume that humans of this era are active members of the Federation of our galaxy..."

Anne: "What's the Federation... I haven't heard you talk about it before?"

Dr. Anderson: "Each galaxy has a Federation or loose-knit organization that includes all sentient life forms on every planet within the galaxy. It would be the equivalent of the United Nations of the galaxy. This Federation has both invited members and observational members. Invited members are those species that have managed to behave in a responsible manner as stewards of their planet and combine both the technology, philosophy, and culture that enable them to communicate as a global entity that has a unified agenda.

"Observational members are species who are fragmented and are still wrestling with one another over land, power, money, culture, and a host of other things that prevent them from forming a unified world government. The human race on planet Earth is such a species, and for now, it is simply observed by the Federation, but is not invited into its policy making and economic systems."

Anne: "Are you saying that our galaxy has a form of government and an economic system?"

Dr. Anderson: "Yes, but if I tell you about this you will lose track of what I really wanted to share with you about the WingMakers..."

Anne: "I'm sorry for taking us off track again. But this is just too amazing to ignore. If there's a Federation of cooperative, intelligent species, why couldn't they take care of these hostile aliens in the year 2011 or at least help us?"

Dr. Anderson: "The Federation doesn't intrude on a species of any kind. It is truly a facilitating force not a governing force with a military presence. That is to say, they will observe and help with suggestions, but they will not intervene on our behalf."

Anne: "Is this like the Prime Directive as it's portrayed on Star Trek?"

Dr. Anderson: "No. It's more like a parent who wants its children to learn how to fend for themselves so they can become greater contributors to the family."

Anne: "But wouldn't a hostile takeover of Earth affect the Federation?"

Dr. Anderson: "Most definitely. But the Federation does not preempt a species' own responsibility for survival and the perpetuation of its genetics. You see, at an atomic level our physical bodies are made quite literally from stars. At a sub-atomic level, our minds are non-physical repositories of a galactic mind. At a sub-sub-atomic level, our souls are non-physical repositories of God or the intelligence that pervades the universe.

"The Federation believes that the human species can defend itself because it is of the stars, galactic mind, and God. If we were unsuccessful, and the hostility spread to other parts of our galaxy, then the Federation would take notice and its members would defend their sovereignty, and this has happened many times. And in this process of defense new technologies arise, new friendships are forged, and new confidence is embedded in the galactic mind.

"That's why the Federation performs as they do."

Anne: "Doesn't BST exist somewhere within the Federation?"

Dr. Anderson: "Yes, it probably does in one of the planets closer to our galactic core."

Anne: "So why doesn't the Federation help... you said they could help didn't you?"

Dr. Anderson: "Yes, they can help. And the Corteum are IMs or invited members and they are helping us. But they themselves do not possess the BST technology... this is a very special technology that is permitted to be acquired by a species that intends to use it only as a defensive weapon. And herein is the challenge."

Anne: "Who does the "permitting "... are you saying the Federation decides when a species is ready to acquire BST?"

Dr. Anderson: "No... I think it has to do with God."

Anne: "I don't know why, but I have a hard time believing that you believe in God."

Dr. Anderson: "Well, I do. And furthermore, so does everyone within the Labyrinth Group—including Fifteen. We've seen far too many evidences of God or a higher intelligence that we can't dispute its existence. It would be impossible to deny based on what we've observed in our laboratories."

Anne: "So God decides when we're ready to responsibly use BST. Do you think he'll decide before 2011?" (I admit there was a tone of sarcasm in this question.)

Dr. Anderson: "You see, Anne, the Labyrinth Group is hopeful that the readiness of the entire species isn't the determining factor, but that a subgroup within the species might be allowed to acquire the technology as long as it was able to protect it from all non-approved forces. This subgroup is hoped to be the Labyrinth Group, and it's one of the reasons why Fifteen has invested so much the ACIO's resource into security systems."

Anne: "You didn't really answer my question though... Do you think it can be developed in 12 years?"

Dr. Anderson: "I don't know. Certainly I hope so, but BST is not our only line of defense. The Labyrinth Group has devised many defensive weapons, not all of which I'll describe to you. The alien race foretold in prophecy is not even aware of Earth at this time. They originate from a different galaxy altogether. The prophecy is that they will send probes to our galaxy and determine that Earth is the best genetic library and natural resource repository in the Milky Way that can be quickly assimilated. They will visit Earth in 2011.

"The prophecy says they will befriend our governments and utilize the United Nations as an ally. They will set about orchestrating a unified world government through the United Nations. And when the first elections are held in 2018, they will overtake the United Nations and rule as the world government. This will be done through trickery and deception.

"I mention these prophecies because they're quite specific as to the dates, and so we have the equivalent of 19 years to produce and deploy BST. Ideally, yes, we'd like to have it completed in order to interface with the intervention points for this race when it decided to crossover into our galaxy. We would like to cause them to choose a different galaxy or abandon their quest altogether. But it may be impossible to determine this intervention point.

"You see, the memory implant technology developed by the Labyrinth Group can be utilized in conjunction with BST. We can define the intervention point when our galaxy was selected as a target to colonize, enter that time and place, and impose a new memory on their leadership to divert them from our galaxy."

Anne: "Either I'm getting tired, or this just got a lot more confusing... You're saying that the Labyrinth Group already has scenarios to nip this thing in the bud... to prevent this marauding group of aliens from even entering our galaxy? How do you know where they are?"

Dr. Anderson: "To answer your question, I would need to explain with much more granularity the precise nature of BST and how it differs from time travel. I'll try to explain it as simply as I can, but it's complex, and you need to let go of some of your preconceived notions of time and space.

"You see... time is not exclusively linear as when it's depicted in a time line. Time is vertical with every moment in existence stacked upon the next and all coinciding with one another. In other words, time is the collective of all moments of all experience simultaneously existing within non-time, which is usually referred to as eternity.

"Vertical time infers that one can select a moment of experience and use time and space as the portal through which they make their selection real. Once the selection is made, time and space become the continuity factor that changes vertical time into horizontal time or conventional time..."

Anne: "You lost me. How is vertical time different from horizontal time?"

Dr. Anderson: "Vertical time has to do with the simultaneous experience of all time, and horizontal time has to do with the continuity of time in linear, moment-by-moment experiences."

Anne: "So you're saying that every experience I've ever had or will ever have exists right now? That the past and future are actually the present, but I'm just too brainwashed to see it?"

Dr. Anderson: "As I said before, this is a complex subject, and I'm afraid that if I spend the time explaining it to you now, we will lose track of more important information like BST. Perhaps if I were to explain the nature of BST, most of your questions would be answered in the process."

Anne: "Okay, then tell me what BST is? Given what the acronym stands for, I assume it means something like... wipe out an event and change the course of history. Right?"

Dr. Anderson: "Let me try to explain it this way. Time travel can be observational in nature. In this regard, the ACIO and other organizations—even individual citizens—have the ability to time travel. But this form of time travel is passive. It's not equivalent to BST. In order to precisely alter the future you have to be able to interact with vertical time, paging through it like a book, until you find the precise page or intervention point relevant to your mission.

"This is where it gets so complex because to interact with vertical time means you will alter the course of horizontal time. And understanding the alterations and their scope and implication requires extremely complex modeling. This is why the Labyrinth Group aligned itself with the Corteum—its computing technology has processing capabilities that are about 3,200 times more powerful than our best supercomputers.

"This enables us to create organic, highly complex scenario models. These models tell us the most probable intervention points once we've gathered the relevant data, and what the most probable outcomes will be if we invoke a specific scenario. Like most complex technologies, BST is a composite technology having five discrete and inter-related technologies.

"The first technology is a specialized form of remote viewing. This is the technology that enables a trained operative to mentally move into vertical time and observe events and even listen to conversations related to an inquiry mode. The operative is invisible to all people within the time they are traveling to, so it's perfectly safe and unobtrusive. The intelligence gained from this technology is used to determine the application of the other four technologies. This is the equivalent of intelligence gathering.

"The second technology that is key to BST is the equivalent of a memory implant. The ACIO refers to this technology as a Memory Restructure Procedure or MRP. MRP is the technology that allows a memory to be precisely eliminated in the horizontal time sequence and a new memory inserted in its place. The new memory is welded to the existing memory structure of the recipient.

"You see, events—small and large—occur from a single thought, which becomes a persistent memory, which in turn, becomes a causal energy center that leads the development and materialization of the thought into reality... into horizontal time. MRP can remove the initial thought and thereby eliminate the persistent memory that causes events to occur.

"The third technology consists of defining the intervention point. In every major decision, there are hundreds if not thousands of intervention points in horizontal time as a thought unfolds and moves through its development phase. However, in vertical time, there is only one intervention point or what we sometimes called the causal seed. In other words, if you can access vertical time intelligence you can identify the intervention point that is the causal seed. This technology identifies the most probable intervention points and ranks their priority. It enables focus of the remaining technologies.

"The fourth technology is related to the third. It's the scenario modeling technology. This technology helps to assess the various intervention points as to their least invasive ripple effects to the recipients. In other words, which intervention point—if applied to a scenario model—produces the desired outcome with the least disruption to unrelated events? The scenario modeling technology is a key element of BST because without it, BST could cause significant disruption to a society or entire species.

"The fifth and most puzzling technology is the interactive time travel technology. The Labyrinth Group has the first four technologies in a ready state waiting for the interactive time travel technology to become operational. This technology requires an operative, or a team of operatives, to be able to physically move into vertical time and be inserted in the precise space and time where the optimal intervention point has been determined. From there the operatives must perform a successful

MRP and return to their original time in order to validate mission success."

Anne: "I've been listening to this explanation and I think I even understand some of it, but it sounds so surreal to me, Dr. Anderson. I'm… I'm at a loss to explain how I'm feeling right now. This is all so strange. It's so big… enormous… I can't believe this is going on somewhere on the same planet that I live. Before this interview, I was worried about balancing my checkbook and when my damn car would ever be fixed… this is just too strange…"

Dr. Anderson: "Maybe we should take another break and warm up our coffee."

Anne: "Signing off for a coffee break…"

(Break for about 10 minutes… Resume interview)

Anne: "If the Labyrinth Group has four of the five technologies ready to go, and is only awaiting the interactive… the interactive part, they must have scenario models and intervention points already established for how they plan to deal with this alien race. Do they?"

Dr. Anderson: "Yes. They have about 40 scenario models and perhaps as many as 5-8 intervention points defined."

Anne: "And if there're that many, there must be a priority established. What's the most probable scenario model?"

Dr. Anderson: "I will be brief on this point because it's such classified information that only the 14x personnel and Fifteen know this. My classification is 12x and so I get diluted reports and quite possibly misinformation with regard to our scenario modeling. About all I can tell you is that we know—from both the prophecies and our remote viewing technology—a significant amount of information about this race.

"For example, we know that it hails from a galaxy that our Hubble telescope has examined as thoroughly as possible and we've charted it as extensively as possible. We know that it is 2.6 million light years away and that the species is a synthetic race—a mixture of genetic creation and technology. It possesses a hive mentality, but individual initiative is still appreciated as long as it is aligned with the explicit objectives of its leaders.

"Because it is a synthetic race, it can be produced in a controlled environment and its population can be increased or decreased depending on the whims of its leaders. It is…"

Anne: "Didn't you just say it's from a galaxy that's 2.6 million light years away? I mean, assuming they were able to travel at the speed of light, it would take them 2.6 million years to come to our planet. And you said earlier that they didn't even know about Earth yet... right?"

Dr. Anderson: "The Corteum come from a planet that is 15,000 light years away, and yet they can come and go between their planet and our planet in the time it takes us to travel to the moon—a mere 250,000 miles away. Time is not linear, nor is space. Space is curved, as your physicists have recently learned, but it can be artificially curved through displacement energy fields that collapse space and the illusion of distance. Light particles do not displace or collapse space, they ride a linear line through space, but there are forms of electromagnetic energy that can modify or collapse space. And this technology makes space travel—even between galaxies—not only possible, but also relatively easy.

Anne: "Why did you say, 'your physicists' just then?"

Dr. Anderson: "I apologize... it's just a part of the conditioning of being isolated from mainstream society. When you operate for 30 years in a secret organization like the Labyrinth Group, you tend to look at your fellow humans... as not your fellow humans, but as something else. The principles of science that the Labyrinth Group has embraced are very different from those taught within your... there I go again... within our universities. I must be getting tired."

Anne: "I didn't mean to criticize you. It's just the way you said it, it sounded as though an alien or an outsider said it."

Dr. Anderson: "I qualify as an outsider, but certainly not an alien."

Anne: "Okay, back to this prophecy or alien race. What do they want? I mean... why travel such a far distance to rule Earth?"

Dr. Anderson: "This seems such a funny question to me. Excuse me for laughing. It's just that humans do not understand how special Earth is. It is truly, as planets are concerned, a special planet. It has such a tremendous bio-diversity and a complex range of ecosystems. Its natural resources are unique and plentiful. It's a genetic library that's the equivalent of a galactic zoo.

"The aliens that are coming desire to own this planet and add it to its colonization plans. As I've already mentioned, this is a synthetic race. A species that can clone itself and fabricate more and more of its population to serve the purpose of its colonization program. However, it desires more diversity, and Earth will represent an opportunity for it to diversify."

Anne: "So where are they right now?"

Dr. Anderson: "I assume they remain in their homeworld... to the best of our knowledge they haven't crossed into our galaxy yet."

Anne: "And when they arrive, how will the ACIO or Labyrinth Group know?"

Dr. Anderson: "As I said, the ACIO has already done a significant amount of intelligence gathering and even selected scenarios and intervention points."

Anne: "So what's the plan?"

Dr. Anderson: "The most logical approach would be to travel to the time and place when the casual thought was born to explore the Milky Way, and through MRP, expunge it from the memory of the race. Essentially, convince them that of all the wonderful, life-inhabited galaxies, the Milky Way is a poor choice. The Labyrinth Group would implant a memory that would lead this race to conclude that our galaxy was not worthy of their serious exploration."

Anne: "So some other galaxy becomes their next target? Wouldn't we bear the responsibility of their next conquest? Aren't we then perpetrators ourselves?"

Dr. Anderson: "This is a fair question, but I'm afraid I don't know the answer."

Anne: "Why couldn't we—using this MRP technology—simply implant a memory not to be aggressive. To tell this race to stop trying to colonize new worlds that aren't theirs to own like property. Why couldn't we do this?"

Dr. Anderson: "Perhaps we will. I don't really know what Fifteen has in mind. I am, though, confident in his approach and its efficacy."

Anne: "But you said earlier that you feared for your life… that Fifteen is probably trying to hunt you down even as we speak. Why are so you confident in his sense of morality?"

Dr. Anderson: "In the case of Fifteen, morality doesn't really play a role. He operates in his own code of ethics, and I don't pretend to understand them all. But I'm quite certain of his mission to avert takeover by this alien race, and I'm equally confident that he will choose the best intervention point with the least influence to the overall species of this alien race. It is the only way he can acquire BST. And he knows this."

Anne: "We're back to God again, aren't we?"

Dr. Anderson: "Yes."

Anne: "So God and Fifteen have this all figured out?"

Dr. Anderson: "There's no certainty if that's what you mean. And there's no alliance between Fifteen and God, at least not that I'm aware of. This is part of the belief system that the Labyrinth Group formalized along the path to developing BST. It's logical to us that God is all-powerful and all knowing because it operates as the universal mind field that interpenetrates all life, all time, all space, all energy… and all existence. This consciousness is impartial, but certainly it's in a position to deny things or, perhaps more accurately, delay their acquisition."

Anne: "If God exists everywhere as you say, then why wouldn't he stop this marauding alien race and keep them in their place?"

Dr. Anderson: "Again, a fair question, but one that I can't answer. I can only tell you that the God I believe in is, as I said before, impartial. Meaning that it allows its creation to express themselves as they desire. At the highest level where God operates, all things have a purpose… even aggressive species that desire to dominate other species and planets. It was Fifteen's belief that God orchestrated nothing but understood everything in the universal mind.

"Remember when I was talking about the galactic mind?"

Anne: "Yes."

Dr. Anderson: "There are planetary minds, solar minds, galactic minds, and a singular universal mind. The universal mind is the mind of God. Each galaxy has a collective consciousness or mind field that is the aggregation of all of the species present within that galaxy. The universal mind creates the initial blueprint for each of the galaxies related to its galactic mind or composite consciousness. This initial blueprint creates the pre-disposition of the genetic code seeded within a galaxy. We, the Labyrinth Group, believed that God designed each galaxy's genetic code with a different set of pre-dispositions or behaviors."

Anne: "And why would this be so?"

Dr. Anderson: "So diversity is amplified across the universe, which in turn permits God to experience the broadest continuum of life."

Anne: "Why is this so important?"

Dr. Anderson: "Because God loves to experiment and devise new ways of experiencing life in all of its dimensions. This may very well be the purpose of the universe."

Anne: "You know you're talking like a preacher? You speak like these are certainties or truths that are just self-evident… but they're just beliefs aren't they?"

Dr. Anderson: "Yes, they're beliefs, but beliefs are important don't you think?"

Anne: "I'm not sure… I mean my beliefs are changing every day. They're not stable or anchored in some deep truth that's constant like bedrock or something."

Dr. Anderson: "Well, that's good… I mean that they change. The Labyrinth Group evolved a very specific set of beliefs—some of these were based on our experiences as a result of the Corteum intelligence enhancement technologies, some were based from ancient texts that were studied, and some were borrowed from our ET contacts."

Anne: "So now you're going to tell me our friendly neighborhood ETs are religious zealots?"

Dr. Anderson: "No... no, I don't mean that they were trying to convert us to their beliefs, we simply asked and they related them to us. Upon hearing them, they seemed quite a bit more like science than religion actually. I think that's the nature of a more evolved species... they finally figure out that science and religion converges into cosmology. That understanding the universe in which we live, also causes us to understand ourselves—which is the purpose of religion and science... or at least should be."

Anne: "Okay, this is getting a little too philosophical for my tastes. Can we return to a question about the WingMakers? If, as you say, there's a galactic Federation that governs the Milky Way, how do the WingMakers factor into this Federation?"

Dr. Anderson: "I'm impressed by the nature of your questions. And I wish I could answer them all, but here again, I don't know the answer. I would assume the Federation and the WingMakers operate in unison and have a mutually beneficial relationship, but I'm not..."

Anne: "But if you can use your remote viewing technology to eavesdrop on this alien race in an entirely different galaxy, why can't you observe the WingMakers and the Federation?"

Dr. Anderson: "Actually, we've tried our remote viewing technology on the WingMakers. It was one of the first things we tried. But we got nothing. In fact, it was the first time when our technology was completely ineffective. We assumed that the WingMakers had developed some form of security that prevented remote viewing. But we weren't sure.

"As for the Federation, they're fully aware of our remote viewing capability, and in fact, we can't eavesdrop on the Federation because they're able to detect our presence if we observe them through remote viewing. So, in deference to their privacy and trusting their agenda, we never imposed our technology on the Federation... perhaps only once or twice."

Anne: "You'll have to forgive me Dr. Anderson, but I find all of this a little hard to believe. We've skimmed the surface of about a hundred different subjects through the course of this interview, and I keep coming back to the same basic issue: Why? Why would the universe be set up this way and no one on Earth know about it? Why all the secrecy? Does someone think we humans are so stupid that we couldn't understand it? And who the hell is this somebody?"

Dr. Anderson: "Unfortunately, there are so many conspiracies to keep this vital information out of the public domain, that what ends up in the hands of the public is diluted to the point of uselessness. I can understand your frustration. I can only tell you that there are people who know about these things, but only Fifteen knows about the larger reality of what we've touched on tonight.

"In other words, and this is to your point, Anne, there are some people within the military, government, secret network, NSA, CIA, etc., that know parts of the whole, but they don't understand the whole. They aren't equipped with the knowledge to stand before the media and explain what's happening. They fear that they would be made to appear feeble by the fact that they only know pieces of what's going on. It's like the story of the three blind men who are all touching different parts of an elephant and each thinks it is something different.

"Fifteen withholds his knowledge from the media and the general public because he doesn't want to be seen as a savior of humanity—the next messiah. And he especially doesn't want to be seen as some fringe lunatic who should be locked up, or worse yet, assassinated because he is so misunderstood. The instant he stepped forward with what he knows, he would lose his privacy and his ability to discover BST. And this he'll never do.

"Most people who know about this greater reality are fearful of stepping into the public scrutiny because of the fear of being ridiculed. You have to admit that the general public is frightened by what it doesn't understand, and they do kill the messenger."

Anne: "But why can't we get even partial truths about this picture of reality... about ETs and the Federation? Someone, the media or government or someone else is keeping this information from us. Like the story you were telling me about the Martians. If this is true and Clinton knows about this, why aren't we being told?"

Dr. Anderson: "There's a cynical part of me that would say something like... why do you watch six hours of television every day? Why do you feed your minds exclusively with the opinions of others? Why do you trust your politicians? Why do you trust your governments? Why do you support the destruction of your ecosystems and the companies and governments that perpetrate this destruction?

"You see, because the whole of humanity allows these things to occur, the wool is pulled over your eyes and it's easy to ration information and direct your attention to mundane affairs like the weather and Hollywood."

Anne: "That's fine for you to say—someone whose IQ can't be charted. But for those of us with average intelligence, what are we supposed to do differently that would give us access to this information... to this larger reality?"

Dr. Anderson: "I don't know. I honestly don't know. I don't pretend to have the answers. But somehow humans need to be more demanding of their governments and even the media. Because the media is a big part of this manipulation, though they're not aware of how they've become pawns of the information cover-up.

"The truth of the matter is that no one entity is to blame. Elitists have always existed since the dawn of man. There have always been those who had more aggression and power and would dominate the weaker of the species. This is the fundamental structure that has bred this condition of information cover-up, and it happens in every sector of society, including religion, government, military, science, academia, and business.

"No one created this playing field to be level and equal for all. It was designed to enable free will and reality selection based on individual preferences. And for those who have the mental capacity to probe into these secrets behind the secrets behind the secrets, they usually find pieces of this larger reality—as you put it. It's not entirely hidden... there are books and individuals and even prophecies that corroborate much of what I've spoken of here tonight. And these are readily available to anyone who wants to understand this larger universe in which we live.

"So, to answer your question: '... what are we supposed to do differently?' I would read and study. I would invest time learning about this larger universe and turn off the television and disconnect from the media. That's what I would do..."

Anne: "Maybe this is a good place to wrap things up. Unless you have anything else you'd like to add."

Dr. Anderson: "Only one thing, and that is that if anyone ever reads this interview, please do so with an empty mind. If you bring a mind full of learning and education and opinion, you'll find so much to argue with in what I've said that you'll not hear anything. And I'm not interested in arguing with anyone. I'm not even that interested in convincing anyone of what I've said. My life will go on even if no one believes me.

"The WingMakers have built a time capsule of their culture and it's magnificent. I wish I could take people to the original site so they could stand before each of the 23 chambers and witness these wall paintings in person. If you were to do this, you would understand that art can be a portal that transports the soul to a different dimension. There is a certain energy that these paintings have that can't be translated in mere photographs. You really need to stand inside these chambers and feel the purposeful nature of this time capsule.

"I think if I could do that, you would believe everything I've said."

Anne: "Could you take someone like me to the site?"

Dr. Anderson: "No. Unfortunately, the security system surrounding this site is so sophisticated, the site, for all intents and purposes, is invisible. All I have are my photographs..."

Anne: "You're saying that if I walked right up to the site, I wouldn't be able to see it?"

Dr. Anderson: "Cloaking technology is not just a science fiction concept. It's been developed for more than 10 years. It's used much more frequently than people realize. And I'm not talking about its diluted version of stealth technology; I'm talking about the ability to superimpose a reality construction over an existing reality that is desired to be hidden.

"For instance, you could walk right up to the entrance of the Ancient Arrow site and see nothing that would look like an entrance or opening. To the observer it would be a flat wall of rock. And it would have all the characteristics of rock—texture, hardness and so forth, but it's actually a reality construction that is superimposed on the mind of the observer. In reality the entrance is there, but it can't be observed because the mind has been duped into the projected reality construction."

Anne: "Great, so there's no way to enter this site and experience this time capsule... so once again, us little humans are prevented from the experience of proof. You see, the reason why this is so hard to believe is that nothing is ever proven!"

Dr. Anderson: "But isn't proof in the eye of the beholder? In other words, what is proof for you may not convince another or vice versa. Isn't this the way of all religions and even science? Scientists claim to have proof of this theory or that theory, and then some years later, another scientist comes along and disproves the previously held theory. And on and on this goes."

Anne: "So what's your point?"

Dr. Anderson: "Proof is not absolute. It's not even objective. And what you're looking for is an experience that is permanent and perfect in its expression of truth. And such an experience, if it indeed exists, is not owned or possessed by any secret network or elitist organization or galactic Federation for that matter.

"You could have this experience of absolute proof tomorrow, and the very next day, doubt would begin to creep in and in a matter of weeks or months this proof or absolute truth that you aspire to possess... it would be just a memory. And probably not even a powerful memory because so much doubt would be infused into it.

"No, I can't give you or anyone absolute proof. I can only tell you what I know to be true for me and try to share it as accurately as I know how with anyone who's interested. I'm less interested in trying to relate the cosmology of the universe than I am in getting the story of the WingMakers and their time capsule into the public attention. The public should know about this story. It's a discovery of unparalleled importance and it should be shared."

Anne: "You do realize don't you, that you've made me the messenger? You've asked me to be the one who takes the public scrutiny and suspicions, and has to endure all of the ridicule..."

Dr. Anderson: "I'm not asking you to do anything against your will, Anne. If you never do anything with the materials I've given you, I'd understand. All I'd ask is that you return them to me if you're not going to get them out. If I step forward as the messenger, I would lose my freedom. If you step forward, this story could catapult your career and you're only doing your job. You're not the messenger, you're the transmitter... the media.

"But you must do what you think best. And I'd understand your decision whatever you decide."

Anne: "Okay, let's wrap it up there. I don't want you to get the wrong impression that I'm a total disbeliever. But I'm a journalist and it's my responsibility to validate and cross check stories before I publish them. And with you, I can't do this. And what you're telling me, if it's true, is the biggest story ever to be told. But I can't take this to the media—at least not the company I work for, because they would never publish it. No validation... no story.

Dr. Anderson: "Yes, I understand. But I've shown you some of the ACIO technologies and photos of the site and its contents, so these must be some form of validation."

Anne: "For me it is, but it doesn't validate all the many claims you've made tonight. For all I know, this Holographic Fractal Object technology you showed me is not so unusual or extraordinary. I'm not a good judge of these things. And even if it were, it certainly doesn't validate the existence of a galactic Federation or the WingMakers for that matter."

Dr. Anderson: "Well... perhaps you're right... we should end this interview. I promised you several interviews before I left. Are we still on for tomorrow night?"

Anne: "Yes."

Dr. Anderson: "Thanks for your interest in my story, Anne... I know it sounds fanciful and outlandish, but at least you've shown restraint in writing me off as a lunatic. And for that, you have my thanks.

"Good night, Anne."

Anne: "Good night. "

8.
Practical Time Travel

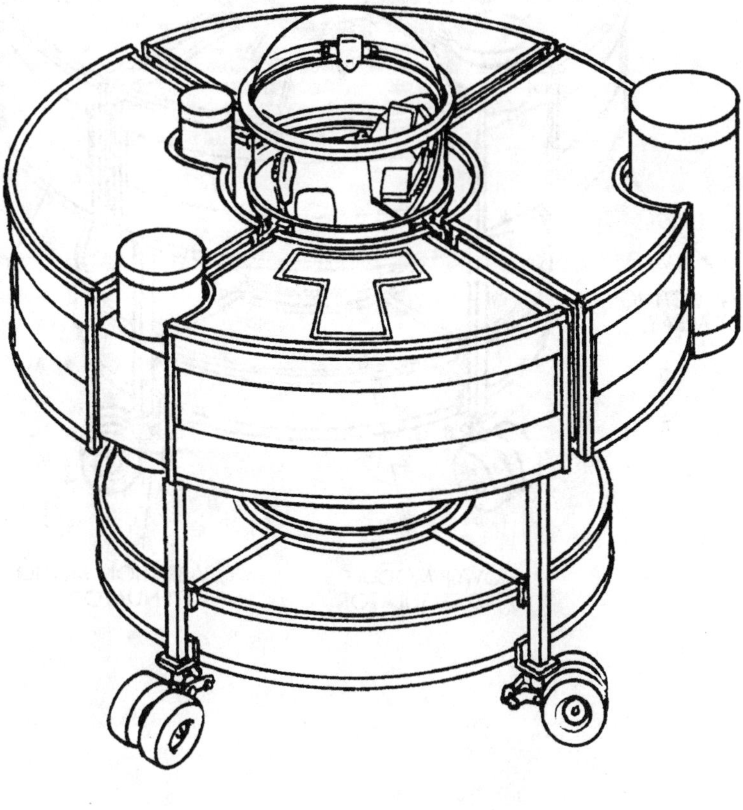

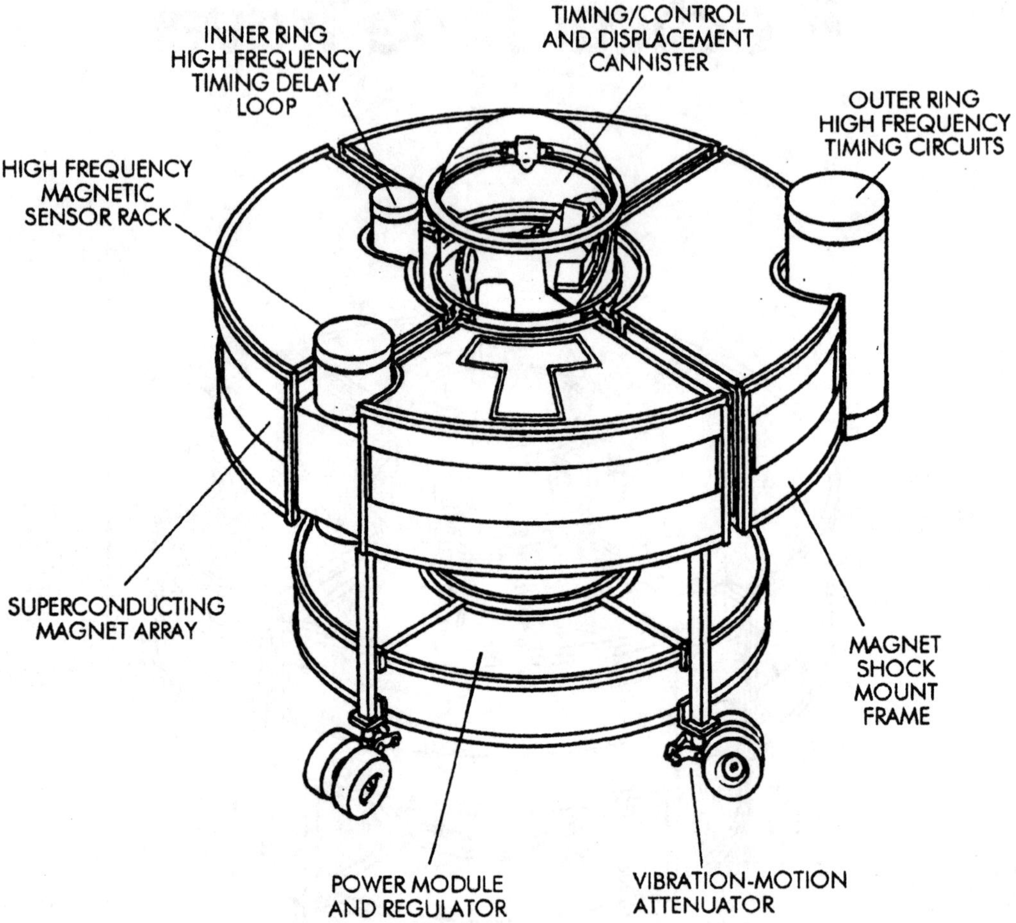
TIMING/CONTROL
AND DISPLACEMENT
CANNISTER
INNER RING
HIGH FREQUENCY
TIMING DELAY
LOOP
OUTER RING
HIGH FREQUENCY
TIMING CIRCUITS
HIGH FREQUENCY
MAGNETIC
SENSOR RACK
SUPERCONDUCTING
MAGNET ARRAY
MAGNET
SHOCK
MOUNT
FRAME
POWER MODULE
AND REGULATOR
VIBRATION-MOTION
ATTENUATOR

8.
Practical Time Travel

Time is God's way of keeping everything
from happening all at once.
—*Time Travel: A Starter's Kit*

"Time's glory is to calm contending kings,
to unmask falsehood and bring truth to light."
—*William Shakespeare*

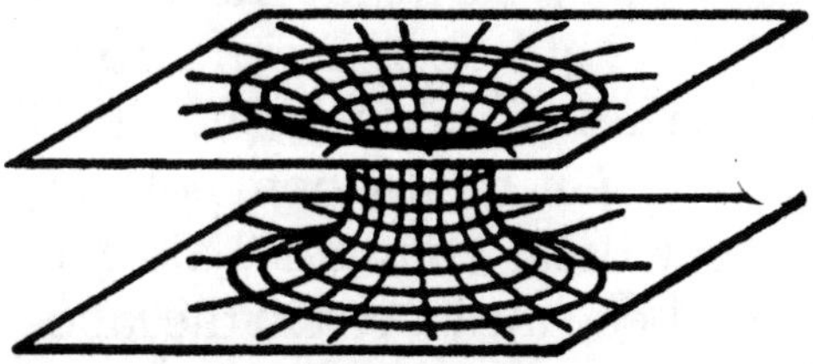

Practical time travel continues to elude the consumer, and until time travel devices are manufactured on a large scale, the garage inventor will just have to build his own.

From the early time travel experiments of Keely, Tesla, the Wilson Brothers of Thorn EMI, and others, to the modern time travel allegations of the Philadelphia Experiment and the Montauk Project, this book hopes to bring practical time travel to the consumer.

Brian X's Views on Time Travel and Interdimensional Voyages

Posted on the Internet was this time travel rant by Brian X. A worthwhile treatise, we include it in this chapter:

Time travel is no longer regarded as strictly science fiction. For years the concept of time travel has been the topic of science fiction novels and movies, and has been pondered by great scientists throughout history. Einstein's theories of general and special relativity can be used to actually prove that time travel is possible. Government research experiments have yielded experimental data that conclusively illustrate that fast moving aircraft have traveled into the future. This phenomenon is due to the principal of time dilation, which states that bodies moving at high velocities experience a time that ticks slower than the time measured at zero velocity.(3) Not as much time elapses for a moving body as does for everything else. Phenomena known as wormholes and closed timelike curves are possible means of time travel into the future and the past.(4) Traveling into the past is a task which is much more difficult than traveling into the future.

Imagine if you will, that you are one of the people still alive today that was born prior to 1903, when the first airplane took flight. When you were young the idea of flying would probably have been quite exciting. Some scientists believe that we may presently be living through an identical scenario. The thing that would be so exciting however, would not be flight, but time travel. Leading scientists believe that our children will live to once again see the impossible become routine. Professor Mikio Mukaku of the University of New York believes that space flight may one day unlock the secret of time itself. This will require the development of spacecraft that can travel at speeds on the order of two hundred million kilometers per second, that's about six hundred million miles per hour. Craft traveling at this speed will take us near the speed of light, where time actually slows down. This is what's known as time dilation. Einstein's theories predict that the faster a spacecraft moves, the slower time ticks inside of it. Imagine that a rocket ship takes off from earth and approaches the speed of light. If we were to watch it from earth with a very powerful telescope as it traveled away from us, we would see everyone inside the ship as being frozen in time. To us their time would slow down, but to them nothing would change! This has been measured in the laboratory and on location using atomic clocks, aircraft, satellites and rockets. It is proven that time slows down the faster you move. In 1975 Professor Carol Allie of the University of Maryland tested Einstein's theory using two synchronized atomic clocks. One clock was loaded on a plane and flown for several hours, while the other clock remained on the ground at the air base. Upon return, the clock on board the plane was found to be ever so slightly

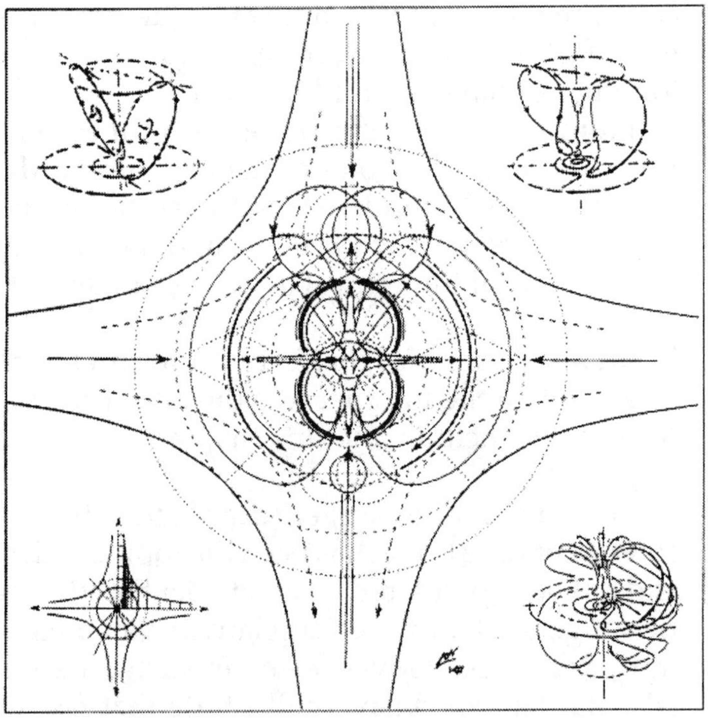

Alfred Wakeman's drawings showing the intertwining continuity of vortex and toroidal inflow/outflow. These 1980 drawings demonstrate dynamic stages of flow development, including time vortices. Courtesy of *Electric Spacecraft Journal.*

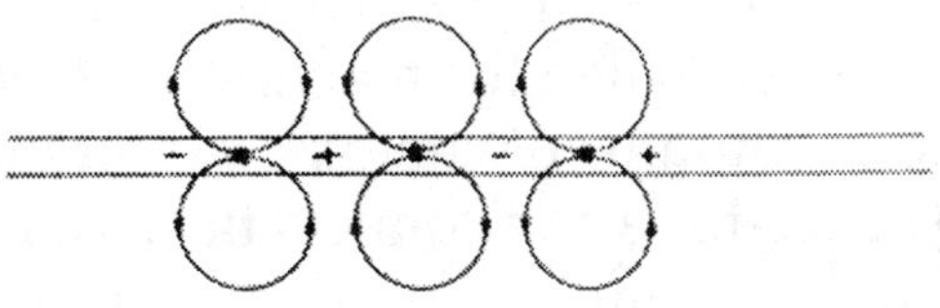

slower than the one on the ground. This was not due to experimental error, and has been repeated numerous times with the same result. This difference in time is even more pronounced in satellites such as the space station. This is because these objects are traveling at speeds much faster and for much longer periods than possible in an airplane. The faster an object moves, the more time is distorted.

Now that we know that it is possible to travel into the future by moving at great speeds, the next problem is how to travel in time a respectable amount without having to sit in a fast moving spaceship for years. This problem is solved by the theoretical existence of what are know as closed timelike curves, and wormholes.

Einstein's special and general theories of relativity combine three-dimensional space with time to form four dimensional space-time.(2)

Space-time consists of points or events that represent a particular place at a particular time. Your entire life thus forms a sort of twisting, turning worm in space time! The tip of the worm's tail would be your birth and its head is the event of your death. The line which this worm creates with its body is called that object's world line. Einstein predicts that world lines can be distorted by massive bodies such as black holes. This is essentially the origin of gravity, remember. Now if an object's world line were to be distorted so much as to form a loop that connected with a point on itself that represented an earlier place and time, it would create a corridor to the past! Picture a loop to loop track that smashes into itself as it comes back around. This closed loop is called a closed timelike curve.

Wormholes are holes in the fabric of four dimensional space-time, that are connected, but which originate at different points in space and at different times. They provide a quick path between two different locations in space and time. This is the four dimensional equivalent of pinching two pieces of a folded sheet of paper together to make contact across the gap. Distortions in space cause the points separated by the gap to bulge out and connect. This forms a wormhole through which something could instantaneously travel to a far away place and time.(4)

No more problems of traveling in a rocket ship for years to get into the future! This is essentially what was written about in "Alice in Wonderland's Through the Looking Glass." Her looking glass was a wormhole that connected her home in Oxford, with wonderland. All she had to do was climb into her looking glass and she would emerge on the other side of forever. In reality however, it would require a much more elaborate scheme to create a wormhole that connects two different points in space-time. First it would require the construction of two identical machines consisting of two huge parallel metal plates that are electrically charged with unbelievable amounts of energy. When the machines are placed in proximity of each other, the enormous amounts of energy—about that of an exploding star—would rip a hole in space-time and connect the two machines via a wormhole. This is possible, and the beginnings of it have been illustrated in the lab by what is known as the Casimir Effect. The next task would be to place one of these machines on a craft that could travel at close to the speed of light. The craft would take one machine on a journey while it was still connected to the one on earth via the wormhole. Now, a simple step into the wormhole would transport you to a different place and a different time.

Wormholes and closed timelike loops appear to be the main ways that time travel into the past would be possible. The limitation on this time travel into the past is that it would be impossible to travel back to a time before the machine was originally created. Although the aforementioned theories of general relativity are consistent for closed timelike curves and wormholes, the theories say nothing about the actual process of traveling through them. Quantum mechanics can be used to model possible scenarios, and yields the probability of each possible output. Quantum mechanics, when used in the context of time travel, has a so-called many-universe interpretation. This was first proposed by Hugh Everett III in 1957.(3) It encompasses the idea that if something can physically happen, it does in some universe. Everett says that our reality is only one of many equally valid universes. There is a collection of universes, called a multiverse. Every multiverse has copies of every person and all matter...

The proposal of time travel is backed by scientific theory, but that is not enough to make it realistically possible. Numerous arguments are proposed that prevent time travel into the past. Both common sense and scientific fact construct serious obstacles. A major argument against time travel into the past is called the autonomy principle, better know as the grandfather paradox. This paradox is created when a time traveler goes back in time to meet his or her grandfather. Now upon their introduction it would be possible to change the course of events that lead up to your grandfather and grandmother marrying. You could tell him something about a family secret to convince him you are who you say you are, and he may proceed to tell his soon to be wife. She may in turn doubt his sanity and have him committed. Thus your grandparents would never have your mother, and therefore you couldn't be born! But then how could you have ever existed to travel back in time if you don't exist?

Another argument of impossibility is called the chronology principle. This principle states that time travelers could bring information to the past that could be used to create new ideas and products. This would involve no creative energy on the part of the "inventor." Imagine that Pablo Ruiz y Picasso, the most influential and successful artist of the 20th century, were to travel back in time to meet his younger self. Assuming he stays in his correct universe, he could give his younger self his portfolio containing copies of his paintings, sculptures, graphic art, and ceramics. The young version of Picasso could then meticulously copy the reproductions, profoundly and irrevocably affecting the future of art. Thus, the reproductions exist because they are copied from the originals, and the originals exist because they are copied from the reproductions. No creative energy would have ever been expended to create the masterpieces!

A notion that was once nothing more than science fiction, is now a concept that's becoming reality. Einstein's theories of general and special relativity can be used to actually prove that time travel is possible, and research has shown that fast moving craft can travel into the future. Time dilation is the easiest method because it merely requires high velocity motion to experience time travel.(3) Phenomena known as wormholes and closed timelike curves are possible means of time travel into the future and the past.(4) Traveling into the past is a task which is much more difficult however. Its theory involves complicated scenarios of tears in four dimensional space-time, energy equivalent to that of an exploding star, and traveling near the speed of light. Both common sense and scientific fact can be used to paint scenarios that become serious obstacles. Yet even these hindrances can be explained away! If the multiverse concept is reality, then most present ideas of time travel are based on a false reality. If time travel is completely impossible then the reason has yet to be discovered.

References

1) Bagnall, Phil , "Where Have All the Time Travelers Gone?" *New Scientist* , July 6, 1996, v. 151

2) Deutsch, David, & Michael Lockwood, "The Quantum Physics of Time Travel," *Scientific American* , March 1994, v. 270

3) Parsons, Paul , "A Warped View of Time Travel," *Science* October 11, 1996, v. 274

4) "How to Murder Your Grandfather and Still Get Born," *The Economist*, January 20, 1996, v. 338

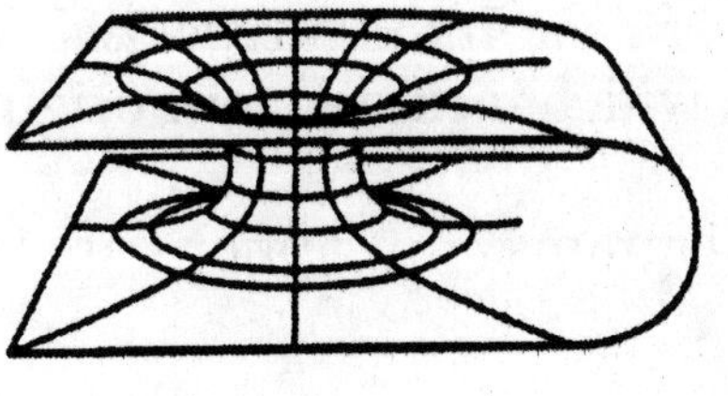

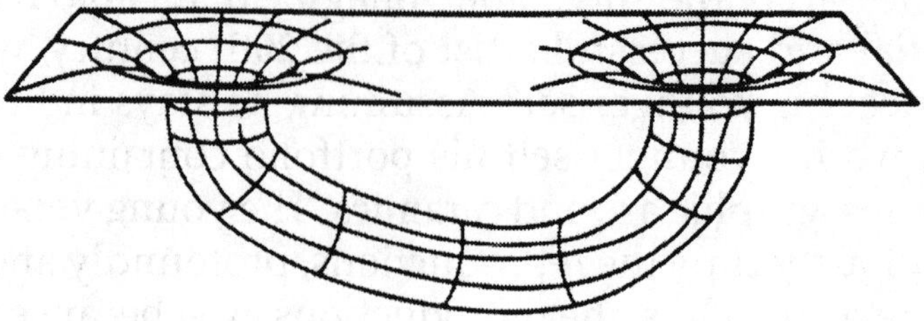

Time Travel Mechanical Services, Ltd.

Now we'll look into another unusual case of time travel, of a group briefly appearing on the Internet. This is the case of Time Travel Mechanical Services, Ltd.

Sometime in December 1998, a man left a folder of documents at the Adventures Unlimited Bookstore and Cafe at 221 Symonds Street, Auckland, New Zealand. The folder was the closed-at-the-ends manila type with the words *Time Travel Mechanical Services, Ltd.* printed across the front.

The man, described as wearing blue jeans and a t-shirt and having his black hair tied in a ponytail, set down the folder and then walked out of the store. On leaving he called out, "With compliments from the future." He never returned.

The New Zealand office did its best to trace the existence of the company, but could find no listing with the New Zealand government (at least in 1998). After examining the documents, they forwarded the originals to the Kempton offices of Adventures Unlimited by special courier.

The contents, a slim file of documents, described what were the plans for a "Polyphase Warp Harmonic Field Array,"—a time travel device widely in use in the 23rd century on a continent known as "Pacifica."

Morrell Chambers III turned the folder over to the nearby Stellar Research Institute where the senior physicist, Winston Whitaker, examined the documents. Whitaker took careful notes on all the diagrams and flew via India and Nepal to Tibet; he hasn't been heard from since. However, certain documents, including the original papers left in Auckland, were still on file in Illinois. This editor was left to work with the original documents and certain notes discovered in a trash can in the Stellar Research Institute's laboratory. Whitaker's experimental geodesic dome suffered a mysterious explosion shortly after his departure for Tibet.

Any discussion of *Time Travel Mechanical Services, Ltd.* should begin with the file documents, which included four original diagrams with brief technical explanations.

The following diagrams and captions are from the original folder.

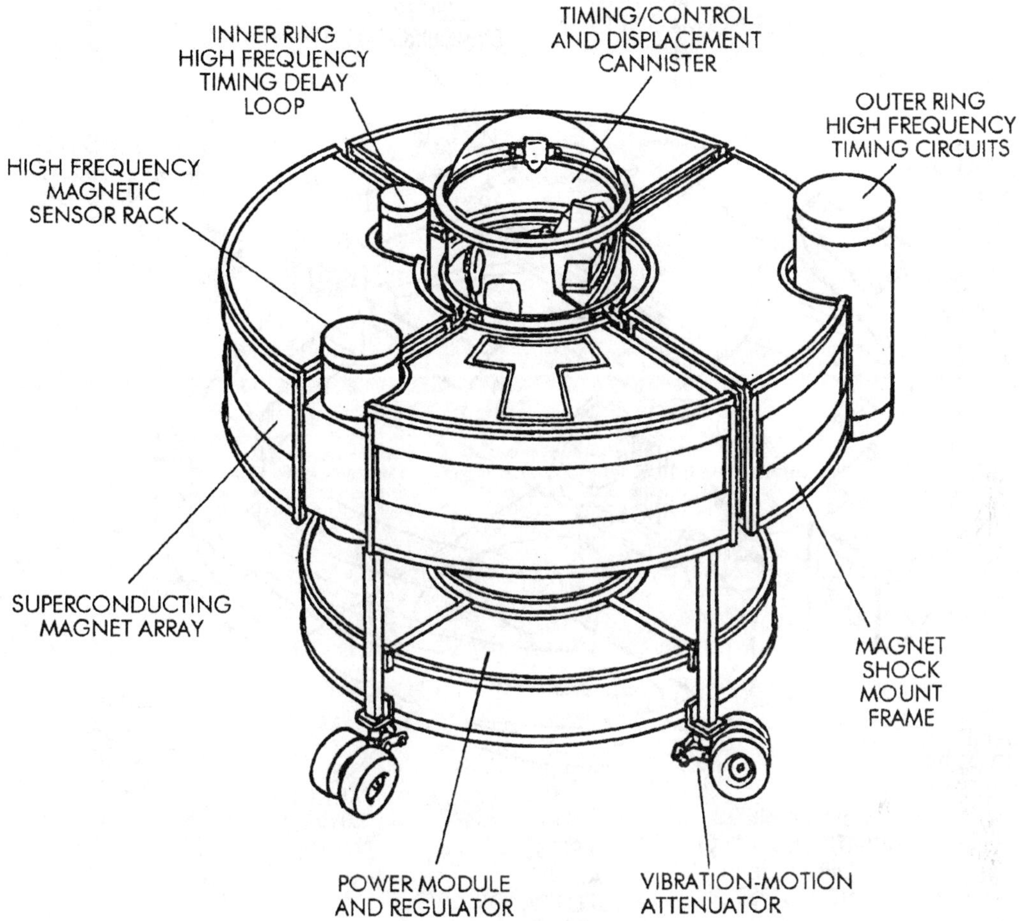

Figure 1

Polyphase Warp Harmonic Field Array

The Polyphase Warp Harmonic Field Array in its complete assembly can be seen in Figure 1. The general arrangement: The radiated portion of a rotating magnetic field in a gigahertz-tetrahertz frequency range lags behind the magnetic generators. This creates unusual harmonic mag field patterns that warp the temporal aspects of a variable region within the magnets. The purpose-built computers generate the special high frequency power pulses and keeps track of the resulting temporal lag.

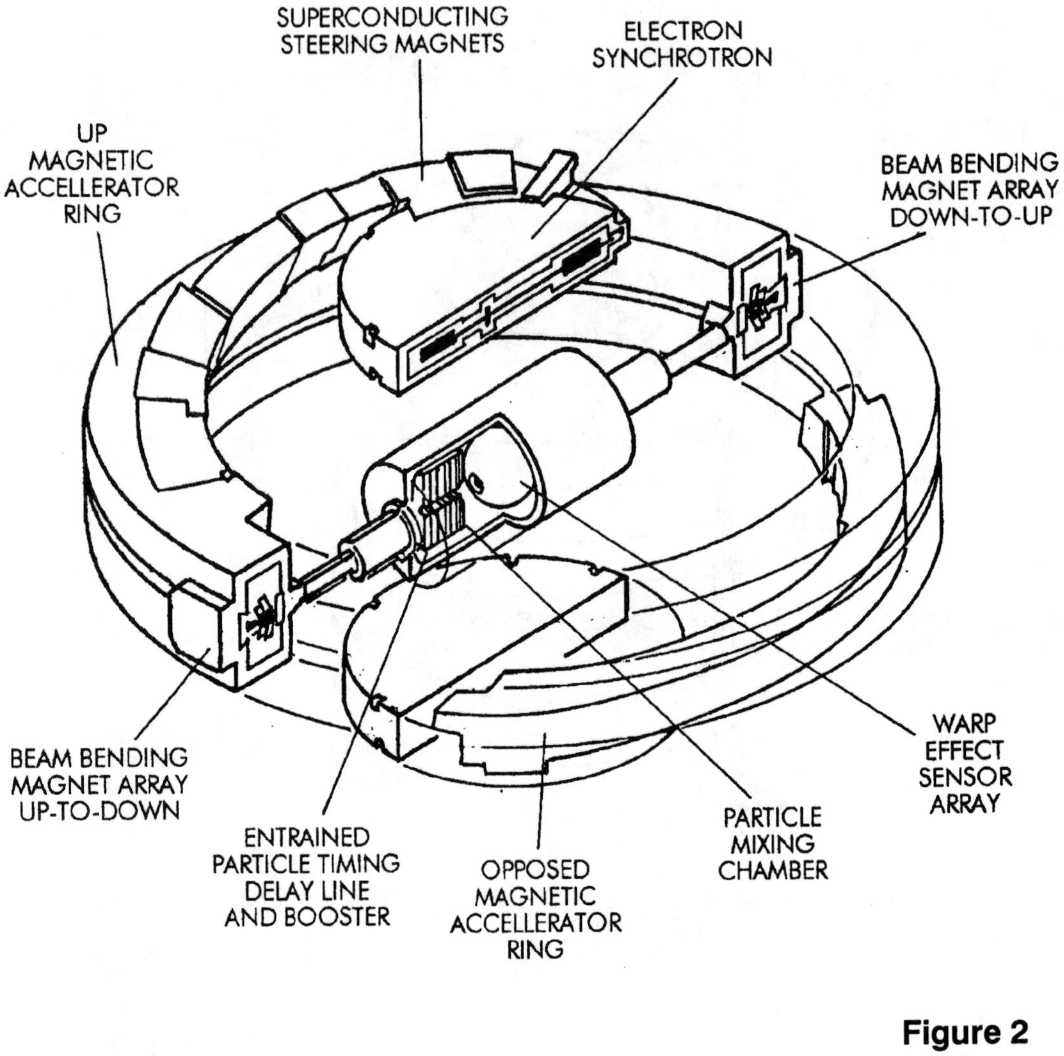

Figure 2

Warp Drive Continuum Booster

The Warp Drive Continuum Booster in its general arrangement can be seen in Figure 2: Counter-rotating streams of electrons are tuned to collide within the particle mixing chamber. The two colliding beams are syncronized to interact in a small region of space where a superimposed pattern of the two signals forms. This small pattern contains regions of temporal continuum disorders which can be measured.

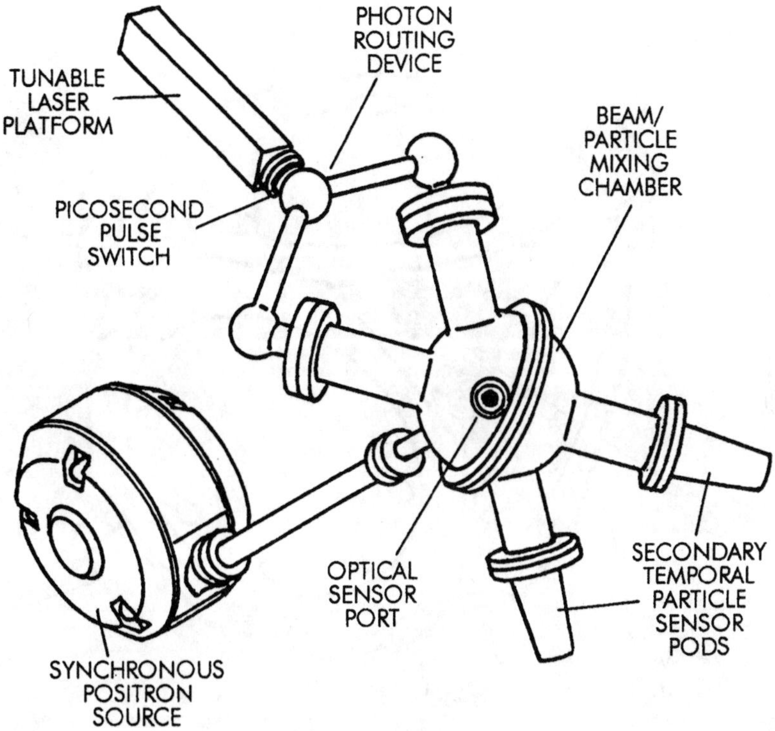

Figure 3

The Time Wave Rectifier

The Time Wave Rectifier in its general arrangement can be seen in Figure 3: Tachyons, daughter products of matter/antimatter collisions, are slowed by magnetic array and tuned laser cooling. The tachyon wavefront can then interact with several replaceable sensors during its transit to the cooling chamber.

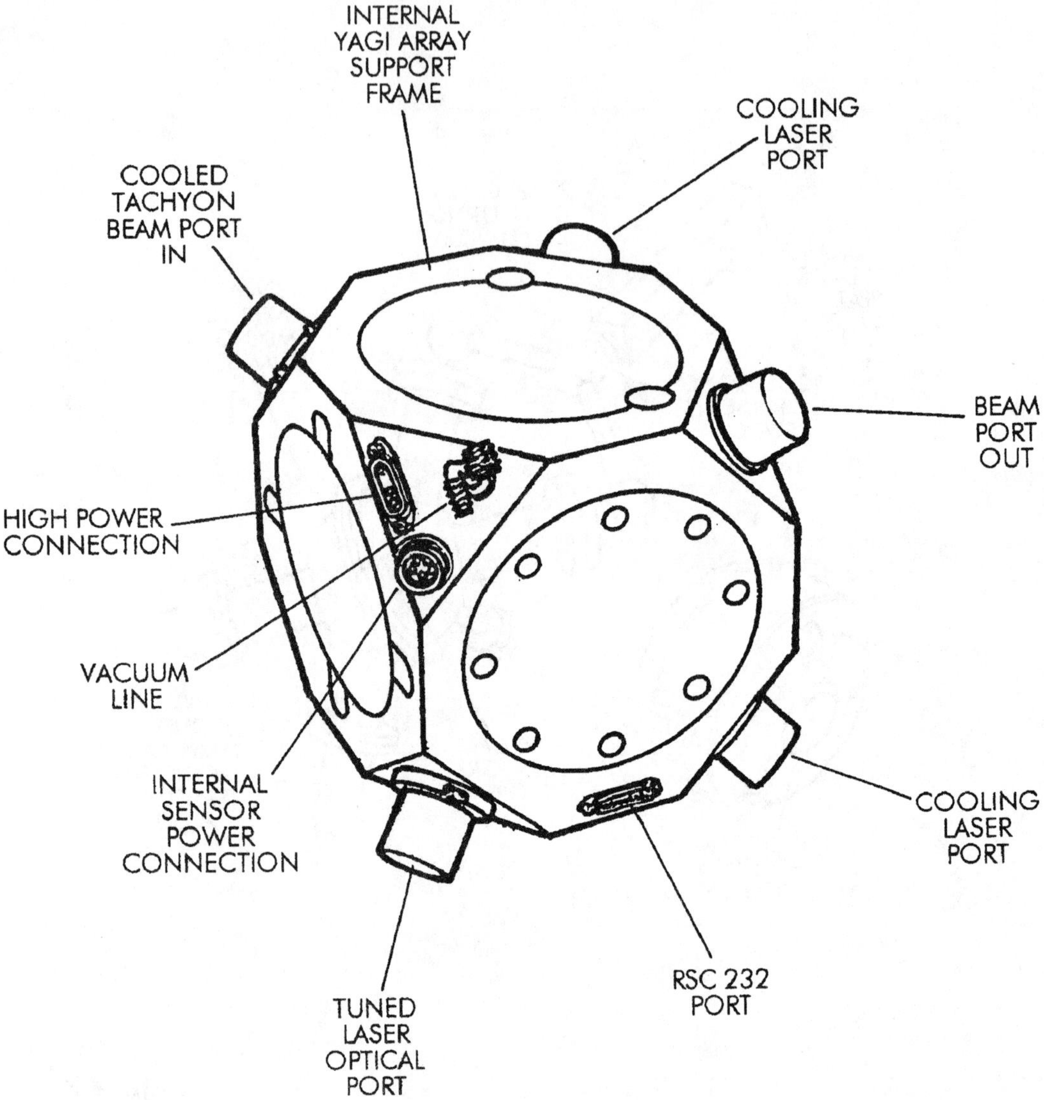

Figure 4

The Zero Time Generator

The Zero Time Generator in its general arrangement can be seen in Figure 4: Specific Energy positrons are timed to interact with specific frequency photons to calibrate "rest" tachyon daughter particle detectors.

The following is taken from the promotional brochure included in the manila folder:

The Polyphase Warp Harmonic Field Array in Action

The Time Wave Rectifier, together with the Zero Time Generator maintain the temporal flux needed to warp forward, or backward, into various hyperdimensional wormholes and various points on the time/space continuum. The Warp Drive Continuum Booster synchronizes the small region of space that is being warped to another space/time continuum. Once a time lock is established through the Zero Time Generator to the destination time and space, the origin, or "home" location, can be terminated, allowing the time/space continuum to snap, like a rubber band, to the selected space/time.

The Polyphase Warp Harmonic Field Array combines the three essential devices for practical time travel and interfaces them in a convenient module that can be installed easily in many discoid or tubular craft.

The Polyphase Warp Harmonic Field Array

The Polyphase Warp Harmonic Field Array in its complete assembly can be seen in Figure 1. The general arrangement: The radiated portion of a rotating magnetic field in a gigahertz-tetrahertz frequency range lags behind the magnetic generators. This creates unusual harmonic mag field patterns that warp the temporal aspects of a variable region within the magnets. The purpose-built computers generate the special high frequency power pulses and keeps track of the resulting temporal lag.

Warp Drive Continuum Booster

The Warp Drive Continuum Booster in its general arrangement can be seen in Figure 2: Counter-rotating streams of electrons are tuned to collide within the particle mixing chamber. The two colliding beams are synchronized to interact in a small region of space where a superimposed pattern of the two signals form. This small pattern contains regions of temporal continuum disorders which can be measured.

The Time Wave Rectifier

The Time Wave Rectifier in its general arrangement can be seen in Figure 3: Tachyons, daughter products of matter/antimatter collisions, are slowed by magnetic array and tuned laser cooling. The tachyon wavefront can then interact with several replaceable sensors during its transit to the cooling chamber.

The Zero Time Generator

The Zero Time Generator in its general arrangement can be seen in Figure 4: Specific energy positrons are timed to interact with specific frequency photons to calibrate "rest" tachyon daughter particle detectors.

According to the original promotional literature found in the folder from the future, the Polyphase Warp Harmonic Field Array, was a complete assembly in itself, and did not require any other devices. With its self-charging, space-power batteries, the full assembly could warp a field in excess of 100 meters. This is enough to take an entire house through space and time. The device could also warp the space/time continuum around cars, ships, submarines, helicopters, discoid "saucers," and tubular or triangular craft. Therefore, many users of The Polyphase Warp Harmonic Field Array's full assembly preferred to have them installed in their vehicle of choice. Some individuals or specialty groups apparently purchased small 5-seater discoid craft from Stellar Aerospace Industries and had the the Polyphase Warp Harmonic Field Array unit installed.

Installation of the unit on the larger SAI 222 craft allowed for larger groups to visit the past, and future. These larger craft, with a capacity of 600 passengers, used a total of five Polyphase Warp Harmonic Field Array units synchronized to the central computer. These ships came into use shortly after the smaller versions, which were known to have cost an exorbitant amount, and were not therefore widely affordable. With the larger craft, affordable commercial traffic would allow for larger groups to visit predetermined time/space destinations allowed by the Temporal Distortion Monitoring Committee set up in 2290.

At least 20,000 of the Polyphase Warp Harmonic Field Array full assemblies were manufactured, and, according to the company brochure, most of them have been scattered throughout time. Their manufacture was discontinued in 2324, due to the popular movement to discontinue further manufacturing on Earth and move all such businesses to other planets and their moons.

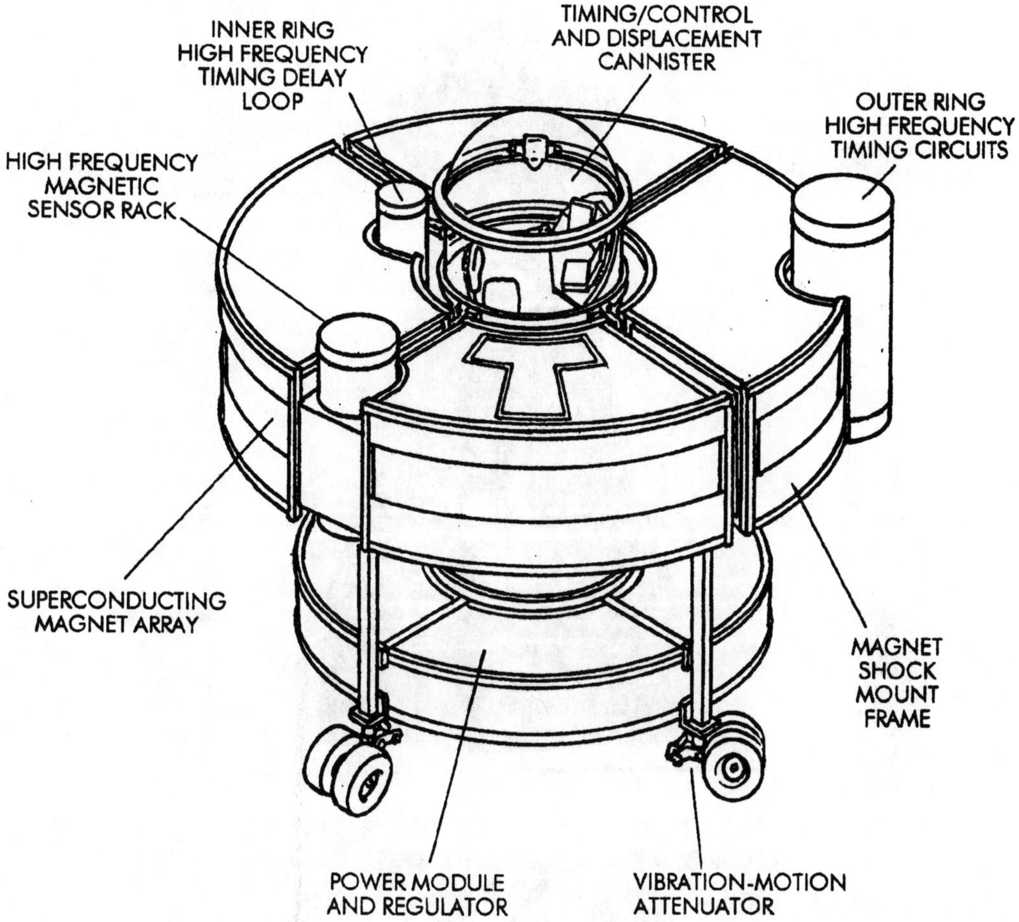
INNER RING
HIGH FREQUENCY
TIMING DELAY
LOOP
TIMING/CONTROL
AND DISPLACEMENT
CANNISTER
OUTER RING
HIGH FREQUENCY
TIMING CIRCUITS
HIGH FREQUENCY
MAGNETIC
SENSOR RACK
SUPERCONDUCTING
MAGNET ARRAY
MAGNET
SHOCK
MOUNT
FRAME
POWER MODULE
AND REGULATOR
VIBRATION-MOTION
ATTENUATOR

TRUE REPORTS OF THE STRANGE & UNKNOWN
FATE
January
1990
USA $1.95
CAN $2.75
U.K £ 2
TESLA'S DEATH RAY
THE MINISTER
OF FAIRYLAND
THE FLYING
HORSEMAN
SEEING YOUR
FUTURE
The New Age & Beyond

9.
Patents & Diagrams

When a distinguished but elderly scientist states
that something is possible, he is almost certainly right.
When he states that something is impossible,
he is probably wrong.
—*Arthur C. Clarke*

Everything that can be invented
has been invented.
—*Charles H. Duell,*
Director of the U.S. Patent Office, 1899

Naturally, time travel devices should be patented, however, this is often a long and costly process. Keeping your device an industrial secret is another way to go.

Time Travel Patents

There are several historical mentions of time travel devices and patents. It was rumored that Bill Lear, the inventor of the Lear Jet, was somehow the owner of a time travel patent.

We have been unable to obtain copies of this patent, however, Stan Deyo reports in his book *The Cosmic Conspiracy,*[13] that Bill Lear mentioned the existence of such an invention. Deyo quotes from a *New York Herald Tribune* article (November 21, 1955) by Ansel E. Talbert:

SPEEDS OF THOUSANDS OF MILES AN HOUR
WITHOUT A JOLT HELD LIKELY

Scientists today regard the Earth as a giant magnet. Many in

> America's aircraft and electronics industries are excited over the possibility of using its magnetic and gravitational fields as a medium of support for amazing 'flying vehicles' which will not depend on the air for lift.
>
> Space ships capable of accelerating in a few seconds to speeds many thousands of miles an hour and making sudden changes of course at these speeds without subjecting their passengers to the so-called 'G-forces' caused by gravity's pull also are envisioned. These concepts are part of a new program to solve the secret of gravity and universal gravitation already in progress in many top scientific laboratories and long-established industrial firms of the nation.
>
> William P. Lear, inventor and chairman of the board of Lear, Inc., one of the nation's largest electronics firms specializing in aviation, for months has been going over new developments and theories relating to gravity with his chief scientists and engineers.
>
> He is convinced that it will be possible to create artificial electro-gravitational fields whose polarity can be controlled to cancel out gravity.

Deyo goes on to say that he saw William Lear on a popular daytime talk show around 1969-1970. Deyo says that the emcee asked Lear what he envisioned the next 20 years of technological advancement would produce. "Lear told him that a person would be able to, say, walk into a New York 'travel' booth—somewhat similar to a telephone box in shape;—deposit his fare; push a button; and walk out the other side of the booth in San Francisco—having been 'teleported' across America in seconds! The studio audience automatically laughed at Lear—much to their uninformed credit. Mr. Lear just gaped at their performance in utter amazement. How painfully sad and lonely he must have felt at the moment when he realized the great gulf that separated the viewing audience from the realities he had already witnessed in the laboratory... He was a kind and sincere man; and this author, for one feels a great loss at Mr. Lear's recent death."[13]

The Maser and Artificial Time Waves

The maser is a device that generates "well organized" or coherent light in the microwave region of the spectrum. Maser is an acronym for microwave amplification by stimulated emission of radiation. In the early 1950s, Charles H. Townes of Columbia University isolated high-energy ammonia molecules using an electric field. The molecules contained electrons raised to high-energy states that then jumped to the ground state, emitting microwave photons, some of which would stimulate more electrons to jump to the ground state. This resulted in stimulated emission, or the spontaneous production of photons with identical wavelength and wave phase. In 1954, Townes, J. P. Gordon, and H. J. Zeiger succeeded in concentrating, or amplifying, such waves, producing the first maser.

When the high-energy molecules (the so-called population inversion) are contained in a partially mirrored chamber, they will strongly interact, emitting an extremely coherent beam from a small opening. Such powerful beams are usually obtained from lasers, the visible-light analogs of masers. Masers are more commonly used in two applications: because maser light is given off in brief pulses at specific frequencies, it serves as the basis for very accurate clocks; and because masers can amplify a microwave signal without generating electrical noise, they find important use as amplifiers of weak microwave signals from distant sources.

Masers could be used to generate artificial time waves and for teleportation. Just as the maser can generate a spontaneous production of photons with identical wavelength and wave phase, so could a maser-type teleportation device be used to create the spontaneous reproduction of an object in another time-space continuum, thus teleporting the object from A to B.

Vortex Drivers for Time Tunneling

Vortex Drivers for time tunneling were said to have come into common usage around 2023, but their use was allegedly keep secret for years. These simple hand-held devices held a battery pack, a miniature warp harmonic field pulse driver coupled with a Whitaker Systems™ vortex driver wave guide.

This inexpensive device looks similar to a small hand-held horn, or perhaps a miniature stereo speaker. It is sometimes worn on the belt, and flatland observers often see a small 'flash' when it is suddenly activated. Observers often have a memory loss, and time will have seemed to have stopped for them.

Because of the small size of the device, it does not have the field range nor computer space for large hyperspace computations. Therefore, it uses an adjustable "time tunneling" device that sends the user, or other person it is directed at, tunneling through hyperspace at a rate varying from 10 minutes to one year. The digital system allows one to program the time length desired and make a sudden leap backward or forward along time's arrow.

The small cone, or vortex driver, sends a pulsed hyperspacial signal into the immediate vicinity of the operator. It can be directed toward the operator, or away from himself at another individual. The dual cone vortex driver allows for both.

For many time travellers, it helps to keep the vortex driver within nearly instant reach, either in the hand, or on the belt. If you are suddenly surprised by someone, or in immediate danger, by activating the device, you can suddenly return to a period 10 minutes before the intruding temporal collision, so to speak.

With a sudden flash of the vortex driver, any witness can be frozen in time, and then sent forward into the future by 10 minutes, or backward. Further time tunneling is then possible, by adjusting the computer. Patents were unavailable at the time, but certain related patents of vortex cones and pulse devices will be shown.

(No Model.) 2 Sheets—Sheet 1.

N. TESLA.

METHOD OF CONVERTING AND DISTRIBUTING ELECTRIC CURRENTS.

No. 382,282. Patented May 1, 1888.

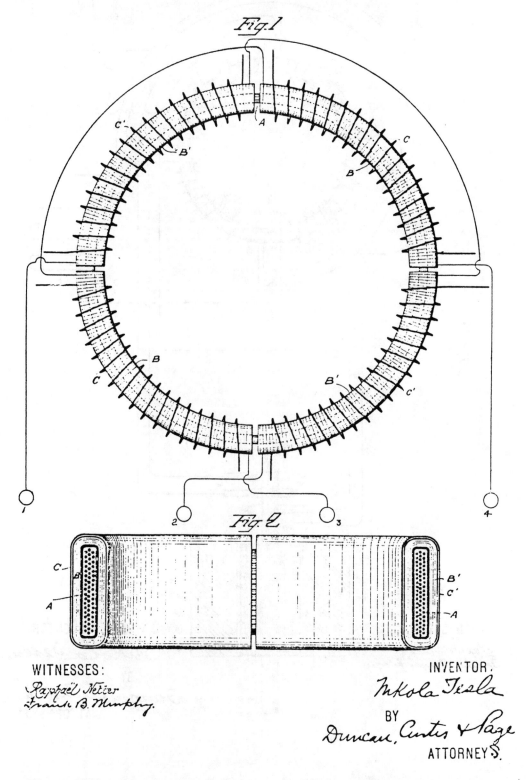

Fig. 1

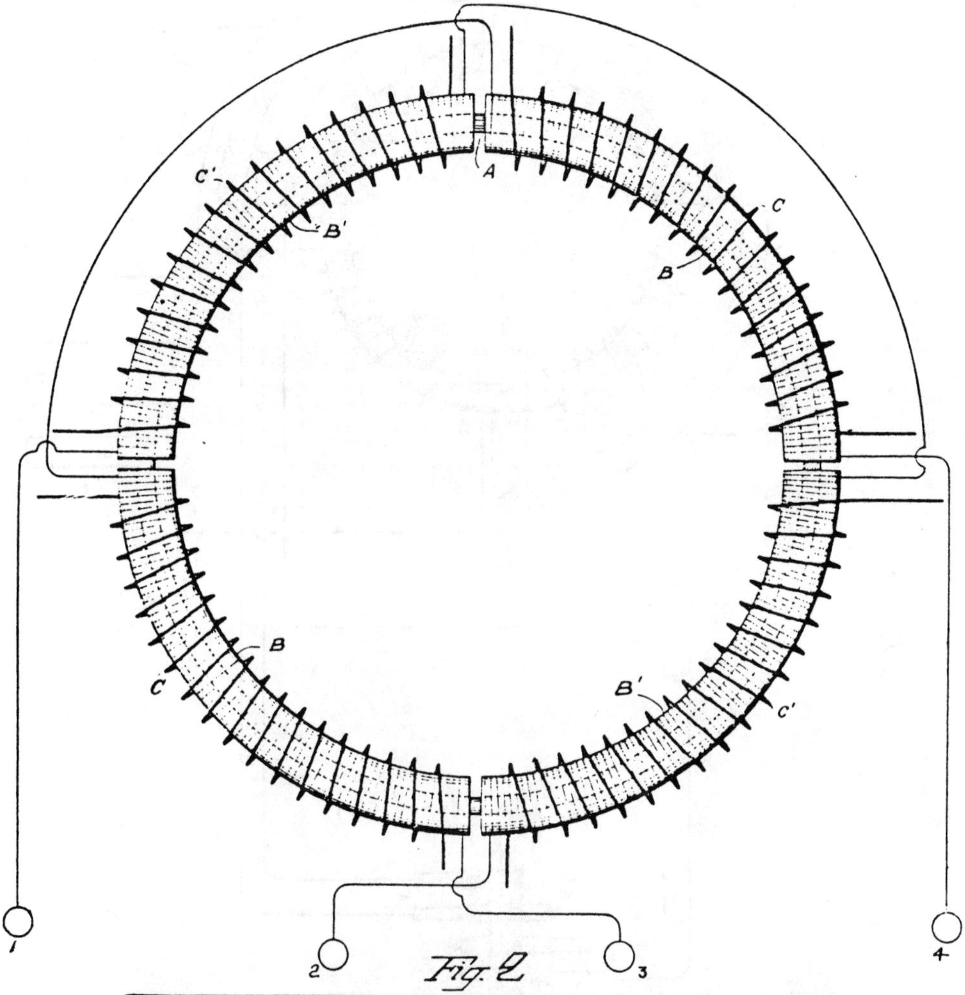

Fig. 2

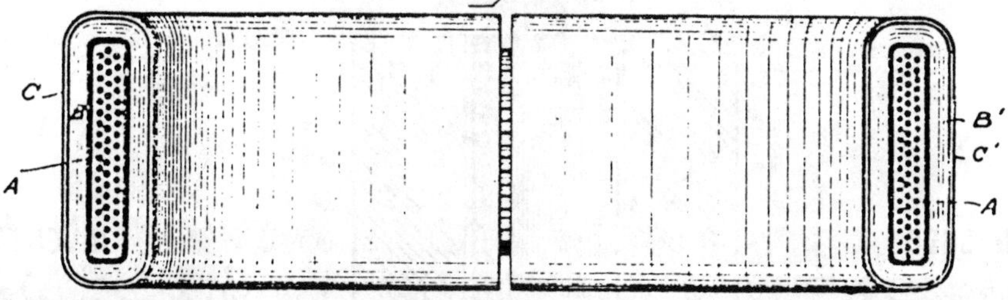

WITNESSES:

Raphaël Netter

Frank B. Murphy

INVENTOR.

Nikola Tesla

BY

Duncan, Curtis & Page

ATTORNEYS.

(No Model.) 2 Sheets—Sheet 2.

N. TESLA

METHOD OF CONVERTING AND DISTRIBUTING ELECTRIC CURRENTS.

No. 382,282. Patented May 1, 1888.

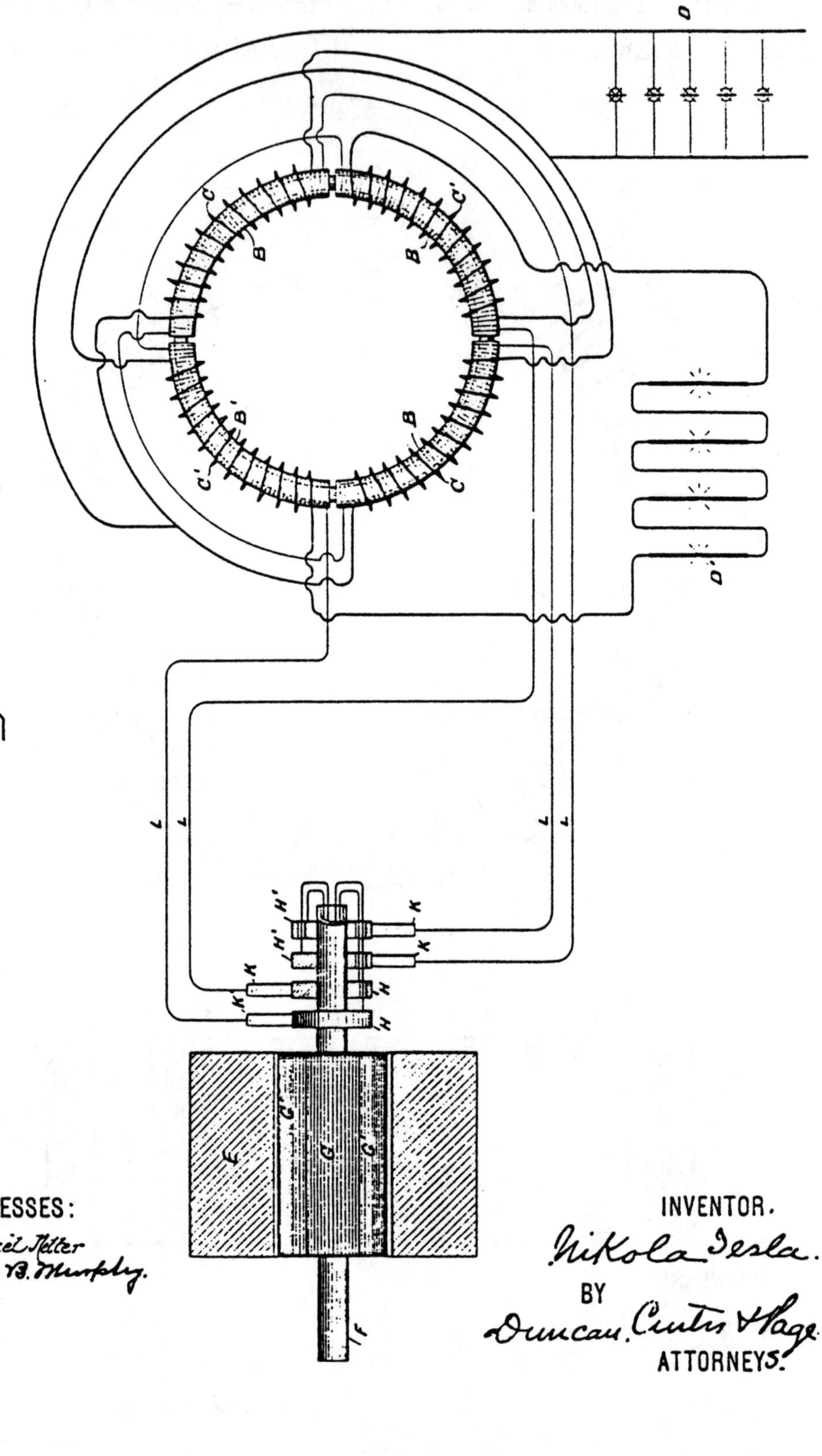

Fig. 3

WITNESSES:

Raphael Netter

Frank B. Murphy.

INVENTOR.

Nikola Tesla.

BY

Duncan, Curtis & Page.

ATTORNEYS.

(No Model.) 2 Sheets—Sheet 2

N. TESLA.

APPARATUS FOR PRODUCING ELECTRICAL CURRENTS OF HIGH FREQUENCY.

No. 568,180. Patented Sept. 22, 1896.

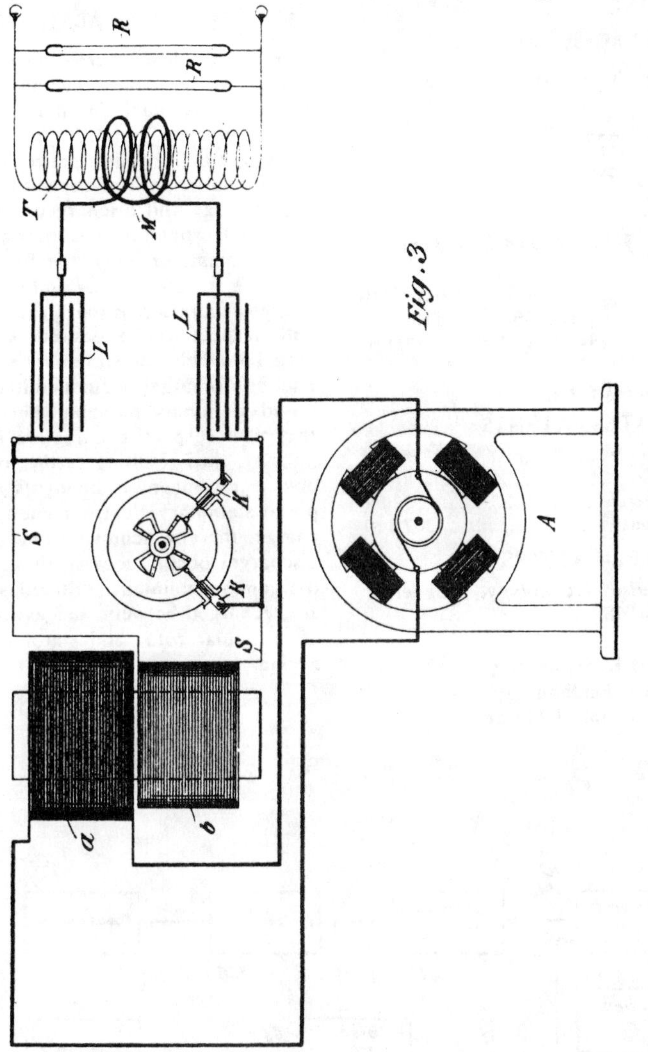

WITNESSES:

Edwin B. Hopkinson.

Benjamin [illegible]

Nikola Tesla INVENTOR

BY

Kerr, Curtis & Page, ATTORNEYS

United States Patent [19]

Gray

[11] **3,890,548**

[45] **June 17, 1975**

[54] **PULSED CAPACITOR DISCHARGE ELECTRIC ENGINE**

[75] Inventor: **Edwin V. Gray**, Northridge, Calif.

[73] Assignee: **Evgray Enterprises, Inc.**, Van Nuys, Calif.

[22] Filed: **Nov. 2, 1973**

[21] Appl. No.: **412,415**

[52] **U.S. Cl.** **318/139**; 318/254; 318/439; 310/46

[51] **Int. Cl.** **H02p 5/00**

[58] **Field of Search** 310/46, 5, 6; 318/194, 318/439, 254, 139; 320/1; 307/110

[56] **References Cited**

UNITED STATES PATENTS

2,085,708	6/1937	Spencer	318/194
2,800,619	7/1957	Brunt	318/194
3,579,074	5/1971	Roberts	320/1
3,619,638	11/1971	Phinney	307/110

OTHER PUBLICATIONS

Frungel, *High Speed Pulse Technology*, Academic Press Inc., 1965, pp. 140–148.

Primary Examiner—Robert K. Schaefer
Assistant Examiner—John J. Feldhaus
Attorney, Agent, or Firm—Gerald L. Price

[57] **ABSTRACT**

There is disclosed herein an electric machine or engine in which a rotor cage having an array of electromagnets is rotatable in an array of electromagnets, or fixed electromagnets are juxtaposed against movable ones. The coils of the electromagnets are connected in the discharge path of capacitors charged to relatively high voltage and discharged through the electromagnetic coils when selected rotor and stator elements are in alignment, or when the fixed electromagnets and movable electromagnets are juxtaposed. The discharge occurs across spark gaps disclosed in alignment with respect to the desired juxtaposition of the selected movable and stationary electromagnets. The capacitor discharges occur simultaneously through juxtaposed stationary movable electromagnets wound so that their respective cores are in magnetic repulsion polarity, thus resulting in the forced motion of movable electromagnetic elements away from the juxtaposed stationary electromagnetic elements at the discharge, thereby achieving motion. In an engine, the discharges occur successively across selected ones of the gaps to maintain continuous rotation. Capacitors are recharged between successive alignment positions of particular rotor and stator electromagnets of the engine.

18 Claims, 19 Drawing Figures

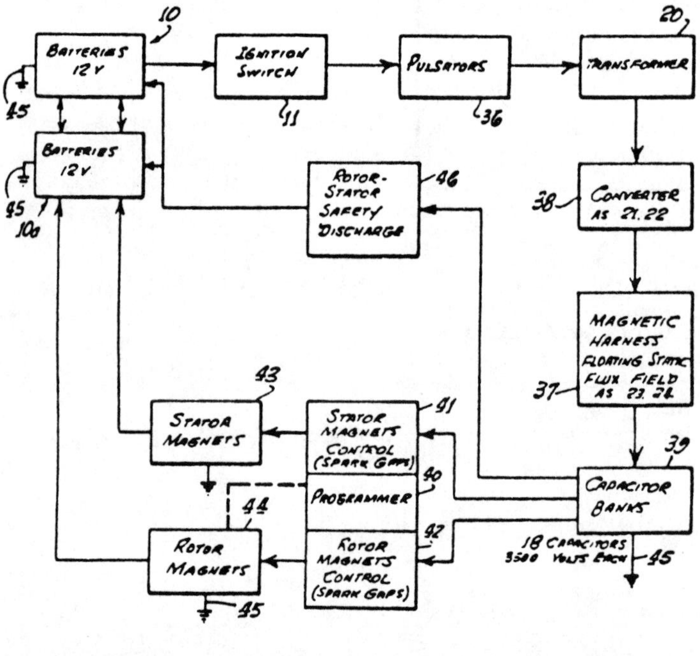

The Pulse Device

by Klaus Schlecht

As a builder, experimenter, and scientific adviser at Dipl.-Ing. Weißmuller, University Karlsruhe, Germany, I would like to thank Charles R. Morton of Lyons, Kansas, for his ideas and dimensions regarding the pulse device.

The experiment was held on January 31, 1985 in the Hochspannungsinstitut of the University Karlsruhe, Germany. Main eyewitnesses were assistant engineer Weißmuller, an electronic engineer, Hoffmann, Karlsruhe and myself.

The pulse device was connected to a circuit, as shown in **Figure 1**. The various test modes and results are listed in **Table 1**. The energy beam, which spreads out during the experiment, was not a conventional version of ion propulsion. This was easily surmised by: the form and the color temperature of the spark inside the spark gap channel; the attraction repulsion behavior of a pair of test balls; and a neon lamp that was 25% ignited (luminous) in the test beam.

The luminous ignition effect would occur if the lamp was held rectangular with their holder--at a distance of about 20 cm to the antenna rod; exactly in the beam spread out direction (circuit mode 1 and 2 in **Table 1**) and between the antenna rod and the lamp. Under this condition, the lamp ignites synchronously with the discharge frequency of more than 1 kHz.

In circuit mode 3 and 4, at 1 meter from the spark gap channel; the suspended test balls--under the attack of the energy beam--began to twist among themselves. This did not apply to the other circuit modes. All test balls, mounted at a 1 meter distance before the antenna rod and the spark gap channel, made a repulsion attraction movement, as a function of the energy wave spread out direction that depended on the circuit mode.

Because of the trafo limit in the power station, it was not possible to tunnel talcum powder with the energy beam of the pulse device into a distant metal box. The amperage gap of the transformer could only lift the powder to a small extent before the fail-safe fuse brought further events to an early end.

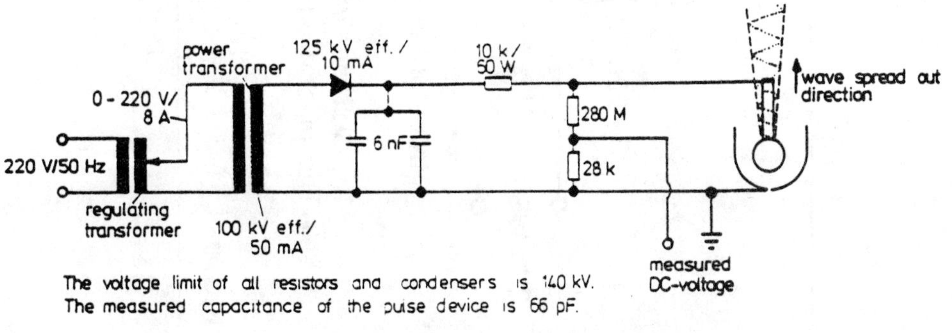

Figure 1 **Circuit Diagram of Connection Mode 1**

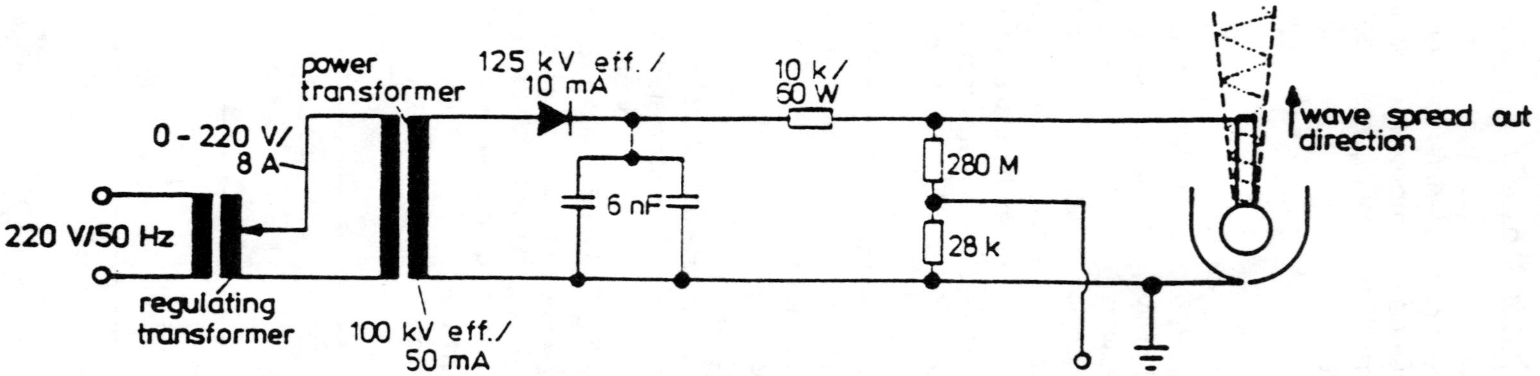

The voltage limit of all resistors and condensers is 140 kV.
The measured capacitance of the pulse device is 66 pF.

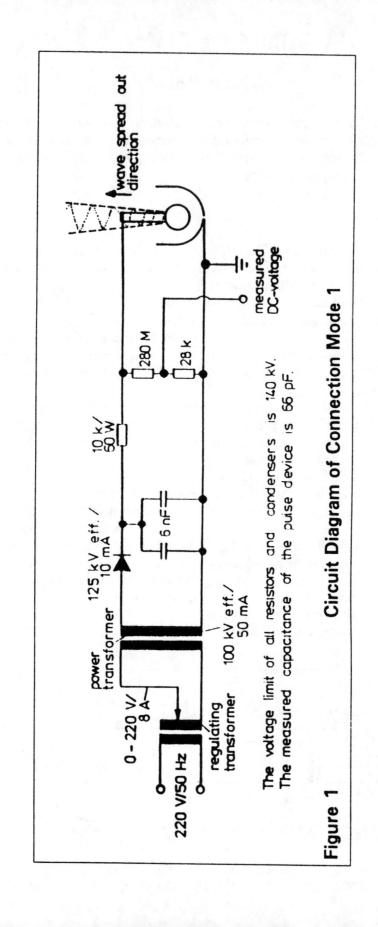

Figure 1 **Circuit Diagram of Connection Mode 1**

mode	spark-gap reduction distance between the hollow valve and the sphere in [cm]	resulting break-down voltage in [kV]	polarity	pulse direction (HF-spread out)	circuit connection
1	0,5	15,5	antenna (tip) +	along antenna (outside)	(wave spread out)
	1,5	32	pulse device (plate) -		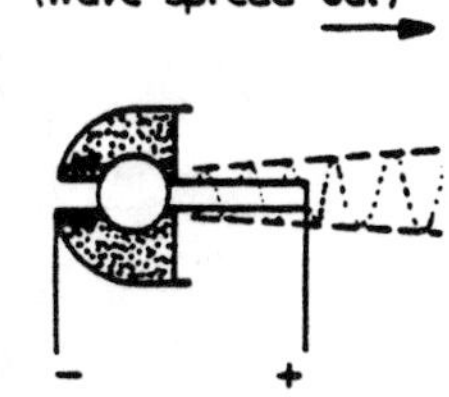
	2,5	43,5			
	3,5	52			
	4,5	58			
	5,5	72			
	6,5	80			
	7,5	89			
	8,5	94			
	9,5	98			
	10,0 (at no reduction gap)	100-106			
2	0,5	17,5	antenna (tip) -	dito.	dito., with opposite polarization
	1,5	30	pulse device (plate) +		
	2,5	43,5			
	3,5	58			
	4,5	78,5			
	5,5	91			
	6,5	101			
	7,5	110			
	8,5	120			
	9,5	122			
	10,0 (at no reduction gap)	120-122			
3	0,5	18	antenna (tip) +	along spark gap channel (outside)	(wave spread out)
	1,5	32	pulse device (plate) -		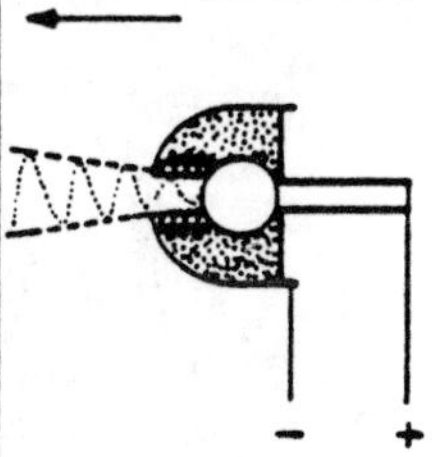
	2,5	45			
	3,5	57			
	4,5	68,5			
	5,5	70,5			
	6,5	79			
	7,5	86			
	8,5	95			
	9,5	97			
	10,0 (at no reduction gap)	99-102			
4	0,5	16	antenna (tip) -	dito.	dito., with opposite polarisation
	1,5	32	pulse device (plate) +		
	2,5	34			
	3,5	48			
	4,5	63			
	5,5	78			
	6,5	100			
	7,5	112			
	8,5	134			
	9,5	(Trafo limit)			
	10,0 (at no reduction gap)	dito.			

Table 1 **Measured DC- values at four connection modes**

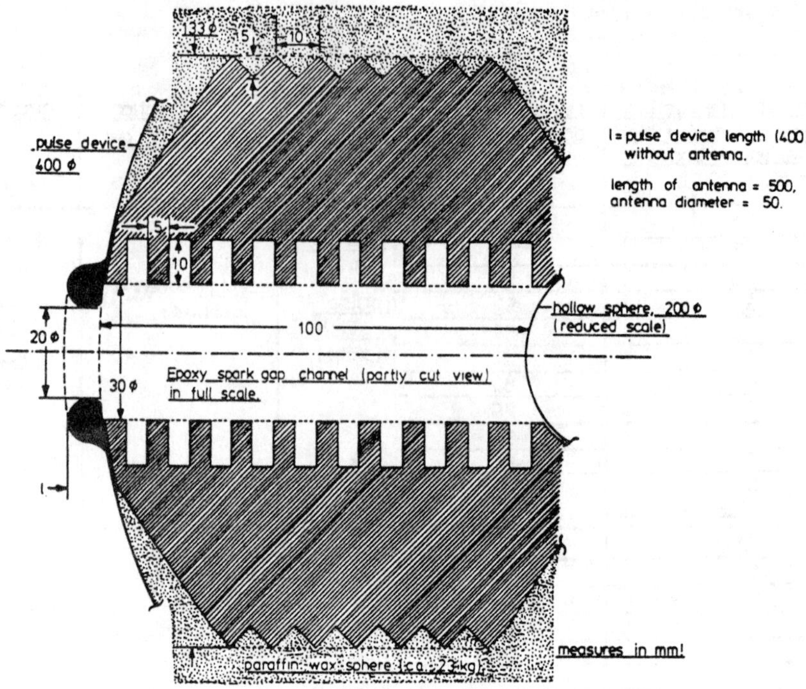

Figure 2 - Spark gap channel

r - Coordinate = radius of the device in 10¹ [mm]. z - Coordinate = position of the earthed plane, which is necessary for reckoning. In 10² [mm].

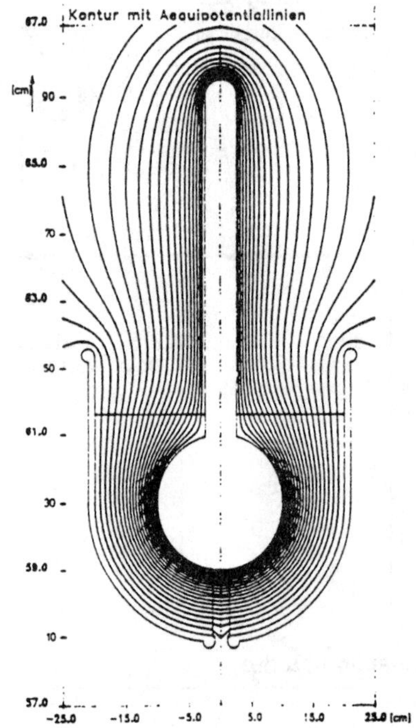

Figure 3 Static field lines of a pulse device with a 200 mm o hollow sphere.

Mathematical model. The charge density has on areas with highest green contrast its maximum. Coordinate explanation see Figure 3. TG 1 = Tachiongenerator Mk. 1.

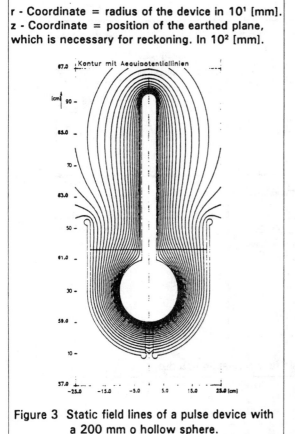

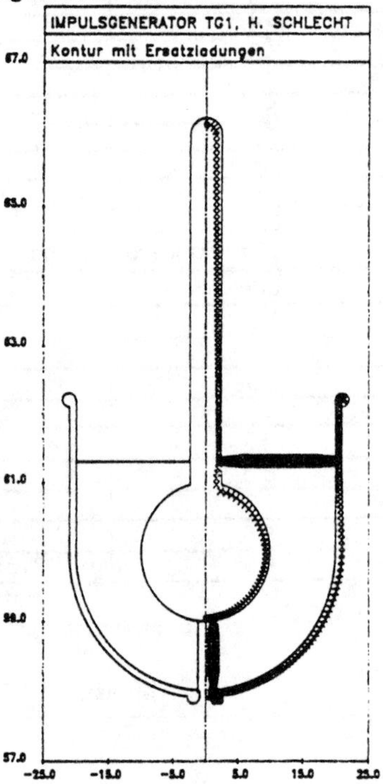

Figure 5 Static charge density of the pulse device from Figure 3

With a Tesla transformer, however, (in the opinion of the eyewitnesses) there will be no lack of energy for a successful powder beaming.

The device has a measured capacitance of 66 pF, designed for a maximum energy storage capacity (see **Figures 3** and **5**). The energy beam spread out direction of the spark gap channel is sufficient, at a distance of about 4 m, to cause air to collapse between the experimenter and the pulse device.

This effect gave rise to fluttering trouser legs and brought a clean and precise spark noise of 100 dB/A minimum sound pressure to our ears.

The discharge frequency could be determined after the given circuit diagram of **Figure 1**:

Discharge Time Constant = R · C
Discharge Time Constant = 10 [kΩ] · 66 [pF]
Discharge Time Constant = 0.66 [ms]

Discharge frequency $\geq$ 1 [kHz]

The resulting pulse device charge (Q) is:

Q = C · U
Q = 66 [pF] · 100 [kV]
Q = 6.6 [coulomb]

ADDENDUM

The electrostatic lifting power, which is necessary to bring a talcum particle in levitation, is about 0.9 pounds.

Therefore at a voltage of 100 kV, for this *given* pulse device, it is estimated that two amperes is necessary in reaching the electrostatic levitation point of a powder particle.

The most efficient power transformer of the university could only deliver 50 mA at 100 kV. Therefore the aforementioned amperage gap =trafo limit.

At 100 kV, the minimum amperage needed to bring the pulse device to work is 300 mA. This value is the upper limit for the fail-safe fuse system.

[*Editor's Note*: Only a small portion of this research report is included in this article. Also, figure numbers represent those in Mr. Schlecht's original.

Additional reference to Charles Morton and his space drive can be found in the article, "Morton's Space Drive," in *ESJ* #4.]

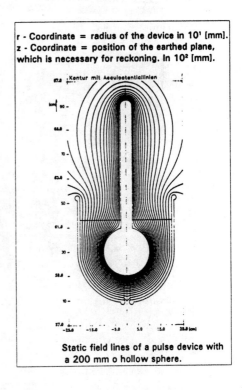

r - Coordinate = radius of the device in 10¹ [mm].
z - Coordinate = position of the earthed plane, which is necessary for reckoning. In 10² [mm].

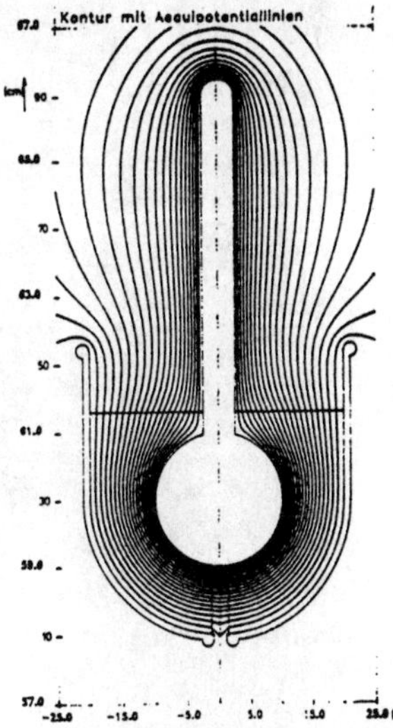

Static field lines of a pulse device with a 200 mm o hollow sphere.

Figure 6 Pulse device - side view

Figure 8 - Pulse device (circuit connection mode 3)

Figure 17 - Dito., inside the cave

Right: Popular radio talk show host Art Bell holds up an alleged time travel device built by Steven Gibbs. Gibbs appeared on Art Bell's show *Coast to Coast* on January 13, 1997. Gibbs claimed on the show that he was a time traveler who had been to the future using his device which he calls a Hyper Dimensional Resonator. Below: A schematic diagram drawn by Gibbs of his Hyper Dimensional Resonator taken from the Art Bell Web Page (artbell.com). Steve Gibbs can be contacted RR #1, Box 79, Clearwater, Nebraska 68726 USA.

THE HYPER DIMENSIONAL RESONATOR

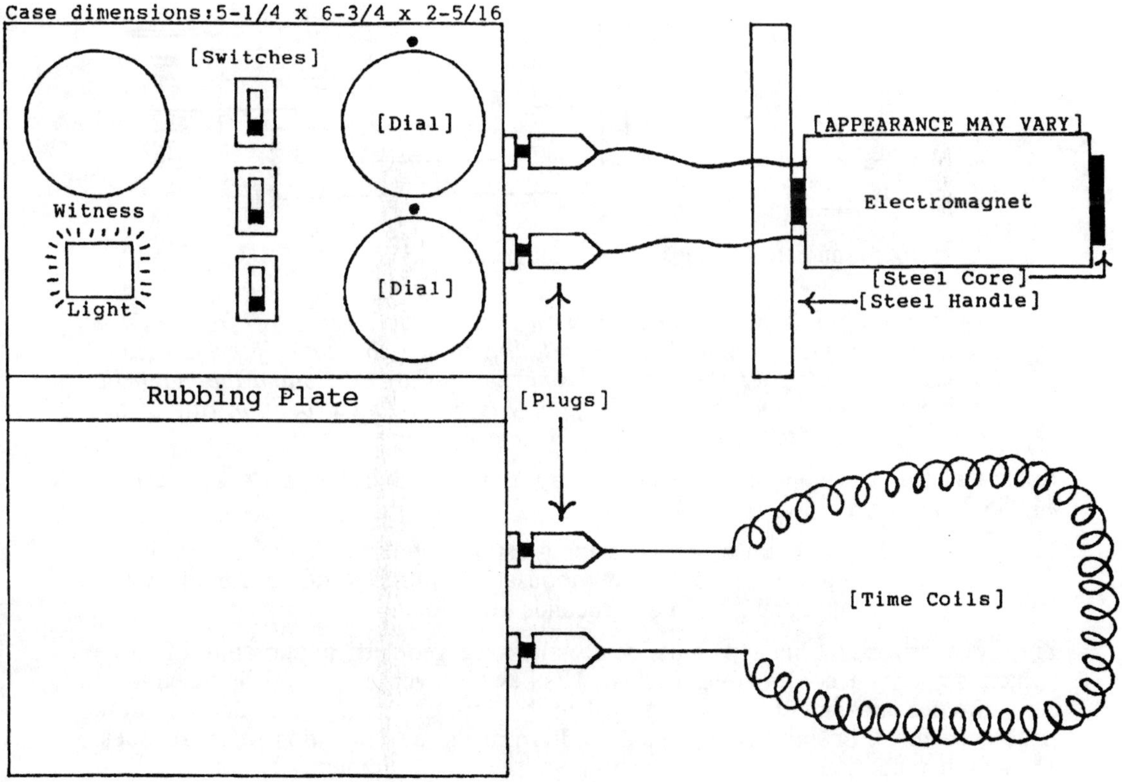

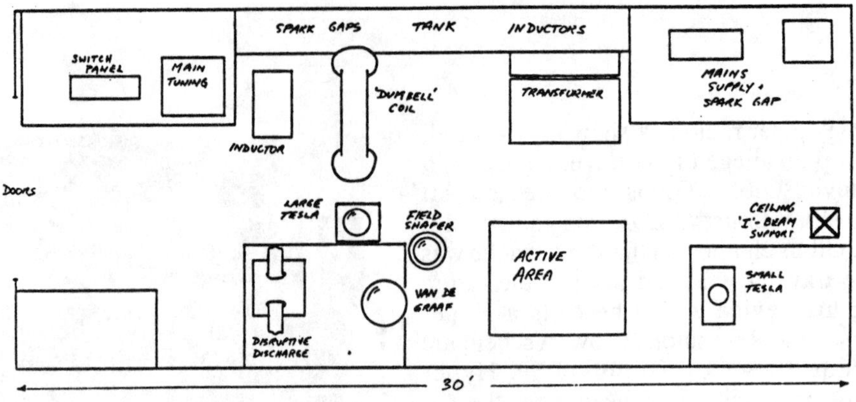

Hutchison Effect - Physical Layout of Experimental Area.

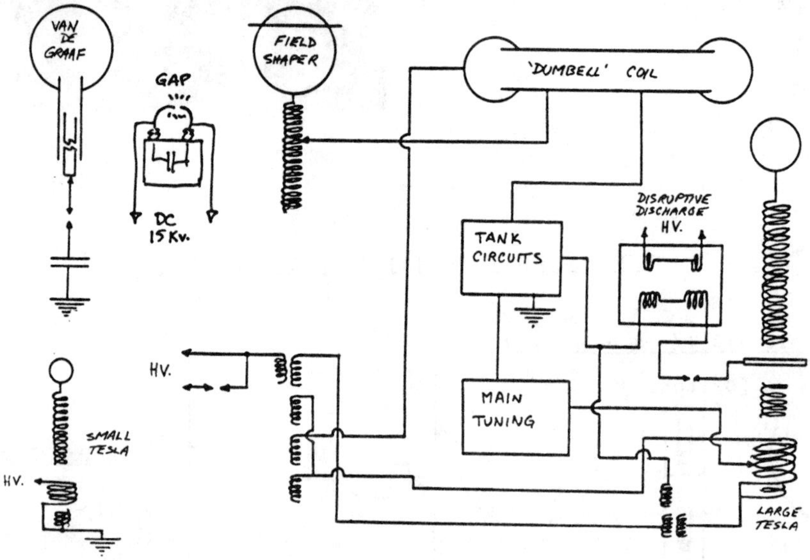

Hutchison Effect - Block Circuit Diagram

> *The traveling scalar trigger-energy has only one degree of freedom for oscillation, namely longitudinal in relation to its direction of propagation, because the vectorial (transversal) properties of the electric and the magnetic fields have been wiped out. Its frequency is that of the repetition rate of the original pulses of the electric and magnetic travelling fields.*

Apparatus for providing the initial pulse trains is found in Zinsser's U.S. Patent #4,085,384 (Apr. 1978).

This "kineto-baric" effect has produced long-lasting impulse trains of several hours duration at peak impulses of up to a few thousand dynes (several oz. force) from a few milliwatts of input energy of a few seconds' duration.

Zinsser theorizes that his apparatus locally alters the geometrical curvature of space (which is gravity according to Einstein). His key discovery is that this induced local curvature does not snap back instantly when the impulse drops to zero and therefore large curvatures, or anisotropies, can result from the accumulation of small ones. Thus high repetition rates of stimulating pulses are desirable.

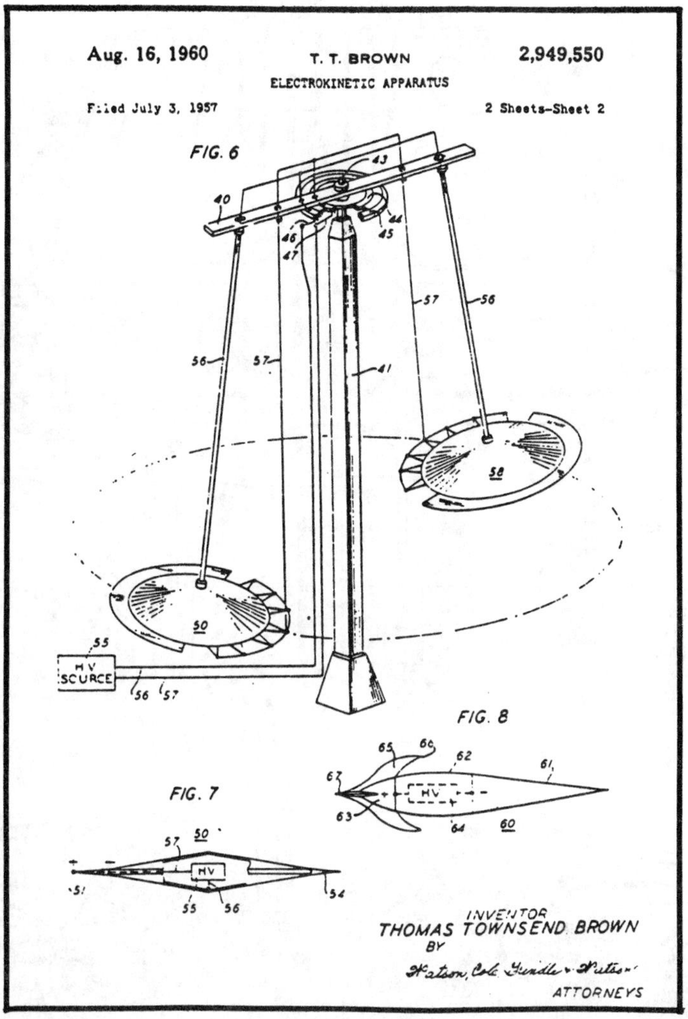

A bona fide technique for producing a local anisotropy in the gravitational field is currently being perfected by Zinsser in West Germany. A proof-of-concept experiment was demonstrated in Toronto in 1981 [59] which showed that, after being stimulated by a particular electromagnetic waveform of short duration, matter activated by the pulses exhibited a definite propulsive thrust due to the localized induced gravitational anisotropy around the matter. The stimulation process is described by Zinsser as follows [60]:

> *A pair of cophasically excited and orthogonally oriented electric and magnetic pulses are made to travel along a parallel conductor system within matter. Suitable elementary dipoles are arranged along the twin conductors. Within the near-field range of the elementary dipoles the fields are superposed on each other in such a way that they are mutually oriented anti-parallel to the "secondary" rotational fields which would form according to the law of induction. From the "primary" moving electric field which is about to separate from an elementary dipole the secondary magnetic rotational field emerges, and vice versa. Typical pulse duration: 2.5 nanosec. or shorter, typical repetition rate: 40 MHz or higher....*

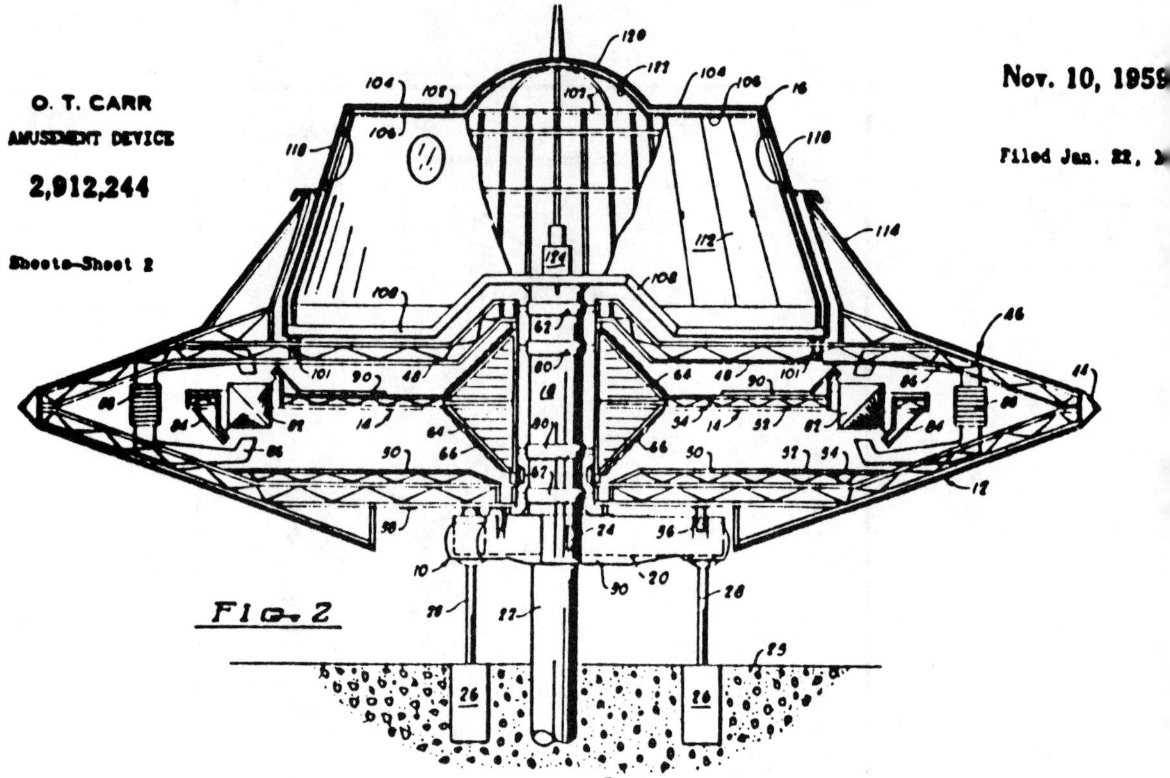

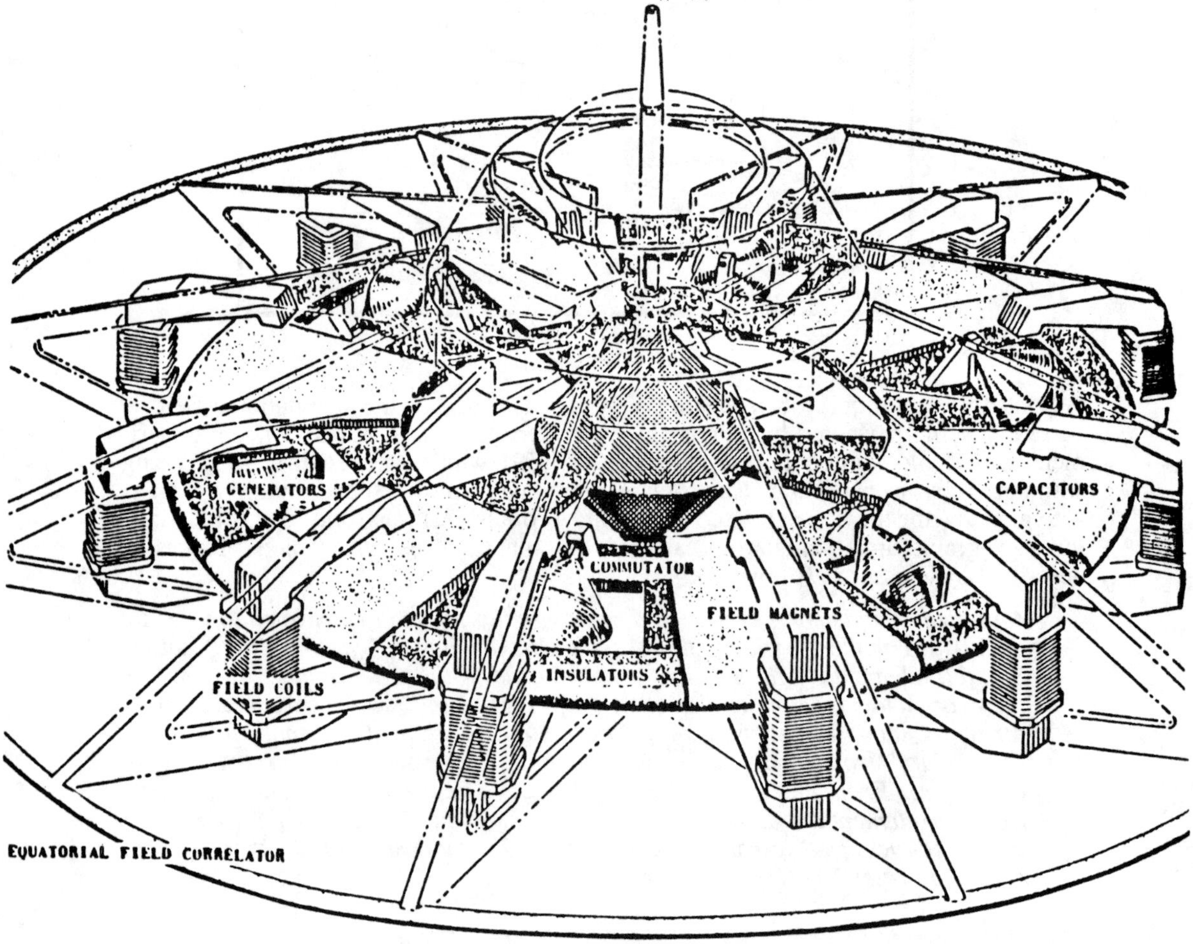

Otis T. Carr's Anti-Gravity Vehicle.

This curious photo was given by an anonymous UFO researcher to author Brad Steiger. Steiger says that the informant claims that this huge piece of unidentified machinery was one of six excavated circa 1990 from a great depth on secret government property somewhere in the U.S. It was allegedly reburied, promptly, in the same area. The machine drarfs the dumpster on the left and the surface is "covered with peculiar hierglyphics that appear similar to characters in ancient Hebrew, Arabic, and Sanskrit.

TIME
MACHINES
SECONDS

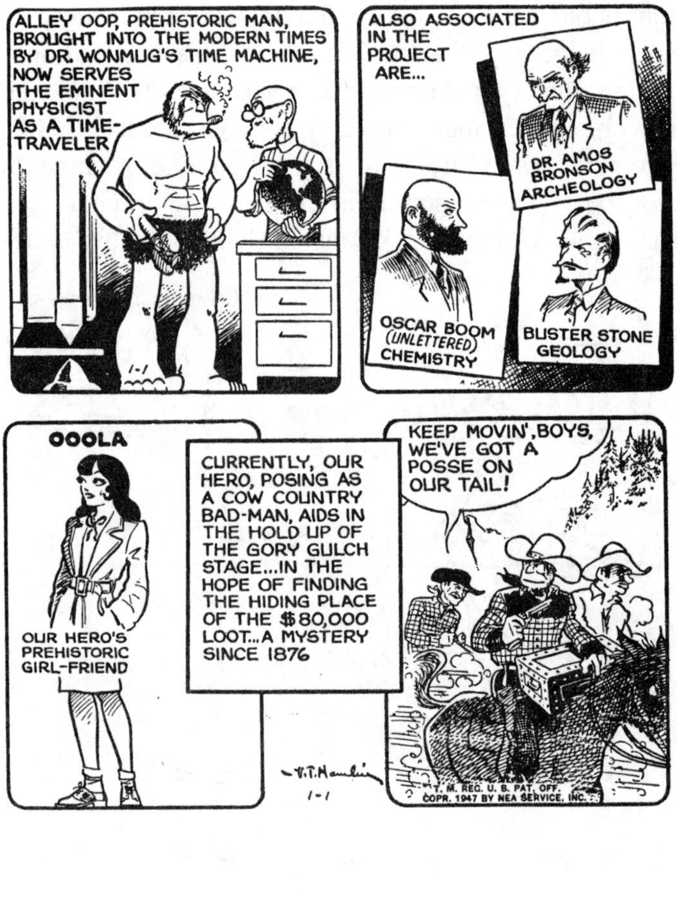
ALLEY OOP, PREHISTORIC MAN, BROUGHT INTO THE MODERN TIMES BY DR. WONMUG'S TIME MACHINE, NOW SERVES THE EMINENT PHYSICIST AS A TIME-TRAVELER
1-1
ALSO ASSOCIATED IN THE PROJECT ARE...
DR. AMOS BRONSON ARCHEOLOGY
OSCAR BOOM (UNLETTERED) CHEMISTRY
BUSTER STONE GEOLOGY
OOOLA
OUR HERO'S PREHISTORIC GIRL-FRIEND
CURRENTLY, OUR HERO, POSING AS A COW COUNTRY BAD-MAN, AIDS IN THE HOLD UP OF THE GORY GULCH STAGE...IN THE HOPE OF FINDING THE HIDING PLACE OF THE $80,000 LOOT...A MYSTERY SINCE 1876
KEEP MOVIN', BOYS, WE'VE GOT A POSSE ON OUR TAIL!
1-1
T. M. REG. U. S. PAT. OFF.
COPR. 1947 BY NEA SERVICE, INC.

Chapter 10.
Time Travel
For Fun and Profit

Isn't it *time* for your machine?
Time Travel is for everyone.
—Stellar Temporal Research
magazine advertisement, March 2009

"I'm going back in time, Phil,
to kill my grandfather."
—Barney to Phil in A Gun for Grandfather (Busby, 1987)

The Time Police

There are a number of devices that time travellers will want to purchase, and plenty of ways to make money while time travelling. This is largely the reason that time travel was briefly made illegal during the years 2056-2068 by the Pacifica Federation in the future. Time travellers began skipping these years, but those who ran the time blockade set up by the so-called "Time Police" were said to have made tons of money. Anyone caught with a so-called "full frequency time modulator" with a zero-time phase transponder was subject to a mandatory prison sentence or possible exile to *Colony 13* on Mars, a fate best avoided, especially during those years.

However, that period is still many years away as far as time's arrow is concerned, and time travel remains a totally legal way to have fun and make a profit as well (but don't go to 2056!).

Knowing the outcome of big money football games or prize fights is an obvious way to make some spending money while time travelling. The world of antiques can also be very profitable for the knowledgeable time traveller. Note that the Highlander of TV fame was an antique dealer, as this is also a good profession for someone whose lifetime spans many centuries or millennia.

Negentropic (Anti-Time) Fields

Speaking of TV, the following article by Jerry W. Decker was posted on the Internet in March of 1998 on KeelyNet (http:\\www.keelynet.com):

I happened to be up late one night and saw a rerun of a *Star Trek Next Generation* (STNG) episode. It was a two parter where the same Enterprise in three different time periods had projected the same kind of beam (inverse tachyon) towards a certain location in space (the Devron system), but from different angles.

This caused a 'rift in time' that caused time to flip and become anti-time, where anything that came into this inverse time field would begin to de-age, to follow a reverse growth path from where it was.

The anti-time effect worked on each mass aggregate as a separate entity from other mass aggregates (a mass aggregate is a combination of many elements to produce a single object, whether that object be a boulder, a chair, a planet, a tree or a human).

I've always found the idea of Time Travel interesting but this Anti-Time idea offered a totally different set of possibilities. In the STNG episode, Enterprise crew members reported old scars healing and they became younger by the day. The blind chief engineer used electronic glasses. He experienced pain due to his eyes healing so that he could once again see.

The episode showed the anti-time field expanding to eventually take over entire star systems, all from this threefold collision of waves that flipped the time stream to initialize the rupture.

Now, the basic idea is pure Tom Bearden (see bibliography) ...using phase conjugation, the same energy is projected from 3 perspectives to collide...the fault to my view is once the beam was shut off in one time period, it would have no further effect on whatever happened in the future.

However, if the time fabric was distorted, then subsequent beams directed to impact the same point but from different angles could cause the anti-time rupture.

The reason I'm writing this is because of the properties of this negentropic field (anti-time).

Entropy is the gradual dissolution of order to a chaotic state. So negentropy is bringing order out of a chaotic state, which is precisely what nature does every day.

Schauberger and others have called it implosion (though Schauberger said a balanced version existed which he called 'impansion')....

There have been reports of activated or energized water that exhibited healing or in some cases rejuvenating properties.

Could there be a way to produce an energy field or zone that would exert 'anti-time' effects on any mass brought into it? Is there a correlation here with reduced or cancelled aether influx into mass?

I think so. Keely said 'Time is Gravity.' If you slow or stop the flow of aether into a mass aggregate, then time slows or stops FOR THAT MASS.

Think of two identical frequencies projected together and displayed on an oscilloscope. By adjusting the phase of one frequency in relation to the other, you can create a standing wave, a bubble or envelope at 180 degrees.

If one frequency is decreased so that the phase difference is less than 180 degrees, the wave will move backward. If one frequency is increased so that the phase difference is greater than 180 degrees, the wave moves forward.

By using this very simple phase conjugation technique, applied to produce a push, pull or balance against aether inflows, we should be able to slow, accelerate or stop time.

Note: Aether influx flowing into the planet traps us to produce what is called 'weight.' If you control the aether influx into one area, anything within that area would be affected by the reduction in aether density. So, we should be able to produce a zone where aether is at least reduced for the zone and thus proportionately for each mass within the zone.

There would be two ways to approach this:

1) to control the aether influx into the earth over a given area OR 2) to control the aether influx ONLY into your OWN body

This functions much like a pressurized cabin, whether that cabin is in an airplane high in the atmosphere where the outside air pressure is less or deep underwater in a submarine where the outside water pressure is much higher than inside the submarine.

The pressure is analogous to aether.

Increase the cabin pressure greater than the outside pressure and it will want to bleed outwards or explode. Anything caught in this high pressure zone would be concentrated or compacted to alter its natural density condition. Over time, this would make the mass aggregation heavier and more dense.

This is the normal condition of mass on a planet because the PLANETARY AETHER INFLUX entrains all masses within its incoming aether flows, thereby making the speed of time on a planet dependent on the density and aether absorption capacity of the planet!!! The same applies to ANY mass aggregate, if aether flows into it, time and gravity is proportionate to the amount of aether inflow into that mass aggregate.

Decrease the cabin pressure so that the outside pressure wants to move inward, either by bleeding inwards or by IMPLOSION. All mass aggregations inside the cabin will be extended or expanded from their natural density condition. Over time, the mass aggregate will become lighter and less dense.

Balance the cabin pressure so that the inside pressure matches the outside pressure and the masses inside will retain their 'steady state' natural density.

It is the basic premise of difference of potential. Nature abhors not only a vacuum but conversely a pressure because it seeks equilibrium through balanced forces.

All of the above is with the understanding that STATIC masses will 'record' their density at the creation of the mass aggregate, and thus absorb a relatively constant flow of aether into it over its span of existence as a mass aggregate.

Static masses aren't truly 'static,' everything ages because of the flow of time that clocks the aether influx into the mass, so it will change much, MUCH more slowly than the more rapidly responsive DYNAMIC mass aggregations, such as living tissue.

DYNAMIC masses will naturally change to match the ambient aether/gravity density or in response to the flow of aether into the mass aggregate (combined) neutral centers.

Dr. Harold Saxton Burr proposed an energy field in the human body which controlled all tissue formation. He called this the 'electrodynamic field.'

Rupert Sheldrake proposed a similar idea which he called a 'morphogenetic field.'

Burr found that all protein in a human body was completely replaced over a six month period. Acomplete record of all the stages of progression of the organism was recorded in 'cellular memory.' This life record is the crux of how anti-time could be used to heal and rejuvenate.

Since living tissues are highly dynamic life processes, then being permeated with an anti-time, reduced aether field could well cause the tissues to rejuvenate to earlier and earlier stages, dependent on the time of exposure to the anti-time field.

In lieu of an energy field, there are numerous reports of rejuvenation using chemical or herbal methods. These appear to act as a catalyst to cause the body to spontaneously rejuvenate on a mass aggregate/systemic level.

Some of the aging antidotes work over a period of hours, others take days to fulfill the rejuvenation. Yet others must be ingested on a regular basis either as food or in a periodic ingestion of the tea, root, herbal mixture or waterlike fluid.

Various rejuvenation stories report:

a) a tea that works only for women will cause them to regain a 20ish youthful condition, yet must be taken on a regular basis, no reported side effects

b) an herbal paste that when taken internally over a period of weeks will cause only women to return to a 20ish condition, take when rejuvenation is needed, no reported side effects

c) a waterlike fluid that when ingested will cause rejuvenation to a late teenage condition over a 12 hour period with no reported side effects, no limitation against men taking the same fluid

d) a tincture of antimony claims to rejuvenate to a late teenage condition over a period of days, side effects include rapid loss of all dead or dying tissue, including hair, eyelashes, eyebrows, finger and toe nails, teeth and an internally generated brackish secretion of corrupt tissue from the skin, works for both men and women, all hair, teeth and nails regrow over a period of 2-3 months

e) a root called 'Amomum' (translated as 'that which makes old men young') as recounted in the Epic of Gilgamesh, said to grow in darkness deep underwater, Patriarchs believed to have eaten it to live 1,000 years or so in full 'vigor,' they aged but retained physical abilities, not the weakening and loss of faculties present in modern times, claim that Noah took samples of it on the Ark to save it from floodwaters, no reported side effects

f) claim that plants which grow in darkness have different life giving and life extending properties, when sought out and eaten exclusively, the subject regains youth, no reported side effects

Of these apocryphal reports, two work only for women, the rest work for both sexes. The main difference between men and women is menstruation. Both sexes have androgen, estrogen and testosterone to various levels, possibly a key. I am of the opinion the menstrual cycle plays a part in why the first two substances work ONLY for women.

One would think that in an anti-time zone one could sleep 8 hours or so which would result in a de-aging effect. Because of 'cellular memory' functioning like magnetic remanence, short duration exposures to the anti-time effect would not be progressively beneficial.

To be most effective, the person desiring full benefit from an anti-time field would have to remain in it for the period of time necessary to attain the desired youthful level.

It is quite probable that time can be accelerated based on increased aether inflows into the mass aggregate. Conversely, anti-time could be accelerated so that healing and rejuvenation would be effected much more rapidly than the normal time flow associated with the planetary body.

The caveat here is there might be a limit on how fast living tissue can be temporally advanced, in either direction.

Thus, you could have an accelerated forward time field that would produce a Dorian Gray type aging effect, where you shrink, wrinkle and lose vitality or life in a matter of minutes (when using the 'normal' time flow as your background timekeeping reference).

By the same token, you could have an anti-time acceleration to reduce an aged adult to a baby (or less) within a matter of minutes, again using the normal planetary flows as the background timekeeping reference.

There are reports of demonstrations given by Indian yogis that appear to involve this anti-time phenomenon. Of course, it could well be hypnosis. However, with consideration of the properties of anti-time, it is well worth recounting one particular demonstration.

A Yogi took a seed, planted it in the earth, watered it and the group of witnesses reported the appearance of a sprout which rapidly grew into a small tree.

Once the tree reached a certain physical size, fruit rapidly appeared and the growth rate stopped. The witnesses reported picking fruit from this tree and eating it as proof of the experience.

The Yogi then resumed his state of concentration, where the tree enfolded into itself, as if a movie of its growth was being run backwards, ending with the bare wet ground. The witnesses still held the half eaten fruit in their hands.

This is an interesting description of just how anti-time would work as described in this paper.

The question remains, how do we control the flow of aether into mass, whether it be into the physical form of a boulder, into a 40 foot circular area on the planet or whether it be for an entire planet?

Once we learn how to control the aether inflows into a mass or control zone, we will be able to produce controlled phenomena at will, ranging from temporal to gravity variations.

Time Travel & the Caduceus Coil

In January of 1998 the following article on time travel and the caduceus coil was posted on the Internet from the Emanon Inventors Association. The story concerns a supposed U.S. Government Temporal Transmission Research Project (known as T.T.R.P.) and their research into the caduceus coil (a.k.a. the tensor coil):

Its main proponent, and inventor, Wilbert Brockhouse Smith developed this coil sometime during the 1950's (Smith died in 1961).

Originally, Dr. Smith was involved in the top-secret program "Project Magnet." This project, allegedly, was a government-sponsored program to produce an aircraft which operated on the same propulsion/flight principles of U.F.O.'s. "Officially," at some point, the program was terminated. At some point, concurrently, Smith was tinkering around with new styles of coil windings. It was during this period of time when he developed the caduceus coil.

The coil is said to produce electromagnetic waves which travel parallel to the coil windings. This violates electromagnetic convention as the standard rule for creating electromagnetism is that the waves are produced perpendicularly (with respect to the coil windings).

Also, supposedly, a caduceus coil transmits these electromagnetic waves at super-luminal velocities (i.e., at faster-than-light speeds).

It is interesting to note that, last year, a Canadian scientist was credited with sending a radio broadcast using "radio-like waves" which were travelling faster than the speed of light.

Smith claimed that timepieces, which were within the caduceus coil's region of greatest magnetic field flux density, measured time at a different "rate and direction of time flow" than other timepieces which were outside of the coil's magnetic field influence.

The caduceus/tensor coil (with proper frequency/voltage) supposedly creates scalar electromagnetic waves that travel at super-luminal velocities (i.e., faster-than-light speeds).

As an electronics engineer, there is one thing that comes to mind regarding this. When an electromagnetic field is created, anything within the densest portion of that field tends to take on the same potential, polarity, frequency, and velocity of that field. This applies both to life forms and inanimate objects.

Some people refer to an object's, or life form's, energy as its "aura," electromagnetic field, or Kirlian field. In fact, there is a seldom known, little used, formula known as the "Johnson Effect" which can be used to ascertain the power levels of an object's, or life form's, electromagnetic field. The reason it was discarded is because it leans towards inaccuracy at frequencies above the Earth's magnetic field frequency (about 7.8 Hz).

With respect to what I've mentioned, I believe Einstein's theory of relativity states that if an electromagnetic wave were to (somehow) be propagated at velocities beyond the speed of light, then that wave would move backward in time.

Or, more specifically stated, that the Universe's "time" would start going backward relative to the "time" experienced by the wave (which, if a wave could have consciousness, would experience a normal, forward, progressive "time").

I believe this is what happened to my friend when he was operating his machine. By the way, he was wearing a digital watch when this happened (the first time). However, the static charge he experienced blew the digital timer circuitry inside the watch. What is really bizarre is I read a story out of the *Weekly World News* (OF ALL PLACES!) that stated the U.S. government was involved in a project called the Temporal Transmission Research Project. The story stated that they place an object inside a hollow aluminum tube and then subject the tube to a high-frequency electromagnetic field that travels faster than light.

In this fashion, supposedly, the tube (and internal contents) are supposed to be able to travel into the past. We all know that aluminum is a great conductor of electricity.

The interesting analogy between that story, and my friend's experience, is that his coils were hollow, were producing extremely high frequencies, and (I believe) were functioning as a caduceus (or tensor) coil.

The *W.W.N.* story also stated that a government scientist had volunteered to go back to the year 1918, remain for 25 minutes of subjective time, then return. They go on to say that he went, but never returned.

Continuing, they said that the project members then decided to look over old publications (from 1918) looking for clues as to what happened. After about two months of searching, they contend, they found an old microfiche copy of a now-defunct police journal which had a very strange article in it.

In the article, it shows what appears to be (in a picture) a metal tube about two feet long and containing the crushed remains of a man. Also, in the picture, is what looks like a cellular phone (lying on the ground) about one to two feet away.

There was a caption below the picture, but I vaguely remember what it said (I was more interested in the picture).

The reason I was so interested in the picture is because of something I noticed about it. I've been working in darkrooms, and with cameras, for more than 30 years and what I saw blew me away.

The picture was taken at night and shows the tube, the "cellular phone" (if that's what it was—suppose it was a "recall device" to allow the man to get back to his own time?), some police officers (in old-style uniforms), some old style police cars, and (in the background) a row of spectators.

The thing that aroused my interest in the picture is that these spectators looked as if they were a good 60-80 feet from the camera. You see, the modern electronic flash (even the ones with the biggest xenon flash lamps) can't illuminate (in darkness) more than about 40-45 feet at the MAXIMUM.

In the old days, the "press 25" flash bulbs (on the other hand) were able to illuminate (in darkness) well up to 80-85 feet or so.

So, it is my contention that MOST of the picture is genuine (that is, taken in 1918). I don't know about the tube and the "cellular phone" though. I can tell you, though, if the picture is a composite (the tube and "cellular phone" being "pasted" into the picture), whoever did it was a master of "special effects / trick photography."

I don't know if the *Weekly World News* would have THAT kind of money as those type of imagery artists are damned expensive to hire. Also, in the picture, I noticed several blades of grass that appeared IN FRONT OF not only the "cellular phone," but the metal tube as well.

The Time Travel Research Center

A search of the world-wide web will result in a number of discoveries, including an interesting organization called the Time Travel Research Center. Their web site can be found at http:\\www. time-travel.com/

What follows is a brief statement of purpose taken from their web site:

Time Travel Research Center ...Explore The Possibilities™

Profile—The Time Travel Research Center, founded in 1995 and based on Long Island, NY, is a privately-owned research laboratory dedicated to the advancement of the science, technology, and research which will someday make it possible for man to travel through time. The Time Travel Research Center's resources are targeted specifically at the ongoing collection, correlation, communication, and development of information and theories in many applied sciences such as physics, mathematics, industrial technology, biotechnology, and other sciences which must be advanced and integrated to make time travel a reality.

From its inception, the Time Travel Research Center has been a leader in the development of capabilities to pursue this goal and is the only company of it kind dedicated exclusively to pursuing the achievement of time travel. Over time, the Time Travel Research Center has continued to grow and expand its efforts and has become a company positioned to remain at the cutting edge of the scientific advances and achievements that may someday help mankind realize time travel capabilities.

Today, the Time Travel Research Center supports private research & development efforts. The Center also pioneered and manages the ongoing development of the TRI-STAR Information System, the world's largest knowledge base of science, technology, and research applicable to the subject of time travel. Utilizing this knowledge base and a world-wide network of scientific information sources the Center continues to advance its research and development at a rapid pace. The Center also founded and manages the Time Travel Research Association, the largest time travel interest group in the world.

Services and Support—The Time Travel Research Center's products and services are targeted specifically towards any efforts which may help advance our progression towards the realization of time travel capabilities. The key focus includes research, development and networking of time travel information and interests from around the world.

Private Research and Development—Research and development of information and theories in applied sciences which must be advanced and integrated to make time travel a reality.

Training Seminars and Consulting—Educational seminars, training program development, consulting, and private and public seminars on many topics relating to time travel technology.

The Time Travel Research™ Journal—A publication from the Time Travel Research Center reviews and explores the latest and most important in scientific advances, theories, and other topics relating to time travel.

Time Travel Research™ Association—The largest time travel interest group in the world founded and managed by the Time Travel Research Center provides many member services and benefits.

Customer Support Services—The Time Travel Research Center attempts to advance the understanding and exploration of this exciting new frontier and the pursuit of time travel through an extensive program of customer support services for Time Travel Research Association members. Services for non-members are also available. Both can be accessed and reviewed via the Center's Internet web site, voice, and fax-on-demand systems.

The Time Travel Research Center created and manages the Tri-Star Information System, the world's largest information and knowledge base of the science, technology, and research applicable to the subject of time travel. The Center continues to use and develop this revolutionary system which goes well beyond concepts of traditional database design and scientific research platforms. Combining knowledge from around the world the Tri-Star System is unmatched in both the quantity and quality of the scientific information it contains. But the system goes further than this by actually modelling and processing complex relationships and "knowledge" vs. simple data. By bridging the walls between physics, mathematics, and computation the Tri-Star system delivers a powerful research and development environment which is actually "tightly coupled" to the actual "knowledge" in the system itself. By linking knowledge and relationships across many sciences and theories the system actually helps us to quickly gain new insight and understanding into many complex relationships and how they work and interact with each other. This is critical to identifying and quickly advancing new scientific relationships that couldn't be seen before through traditional approaches. This sophisticated new tool, the Tri-Star Information System, has been designed with one goal in mind—advancing our efforts to achieve time travel technology.

The greatest scientific quest in history is underway and leading to discovery that will change our world and lives in ways we can't even begin to comprehend. Join this scientific quest and bring the pursuit of time travel directly to you with a free membership or charter membership in the Time Travel Research Association.

Time Travel Research Association members are part of a unique and elite world-wide group experiencing the excitement, discovery and adventure as the pursuit of time travel races on. And you too can explore the possibilities.

The Time Travel Research™ Association is the largest time travel interest group in the world with a mission and direction that has remained intense, daring, and generated a global reach and attention. As a Time Travel Research Association member you can be provided with access to a wide range of information and services—including extraordinary reviews of past achievements, latest breakthroughs, and important developments in the areas of science that address or might affect the ability of man to travel through time.

Teleportation at the California Institute of Technology

The following news story was released in the fall of 1998:

WASHINGTON (Reuters)—They may not be able to ask Scotty to beam them up yet, but California researchers said Thursday they had completed the first "full" teleportation experiment.

They said they had teleported a beam of light across a laboratory bench. They did not physically transport the beam itself, but transmitted its properties to another beam, creating a replica of the first beam.

"We claim this is the first bona fide teleportation," Jeff Kimble, a physics professor at the California Institute of Technology, said in a telephone interview.

Kimble thinks the experiment shows quantum teleportation can eventually transform everyday life.

Scientists hope that quantum computers, which move information about in this way rather than by using wires and silicon chips, will be infinitely faster and more powerful than present-day computers.

"I believe that quantum information is going to be really important for our society, not in five years or 10 years, but if we look into the 100-year time frame it's hard to imagine that advanced societies don't use quantum information," Kimble said.

"The appetite of society is so voracious for the moving and processing of information that it will be driven to exploit even the crazy realm of quantum physics."

Quantum teleportation allows information to be transmitted at the speed of light—the fastest speed possible—without being slowed down by wires or cables.

The experiment depends on a property known as entanglement—what Albert Einstein once described as "spooky action at a distance."

It is a property of atomic particles that mystifies even physicists. Sometimes two particles that are a very long distance apart are nonetheless somehow twinned, with the properties of one affecting the other.

"Entanglement means if you tickle one the other one laughs," Kimble said.

In the weird world of quantum physics, where the normal ideas of what is solid or what is real do not apply, scientists can use these properties to their advantage.

What Kimble's team did was create two entangled light beams—streams of photons. Photons, the basic unit of light, sometimes act like particles and sometimes like waves.

They used these two entangled beams to carry information about the quantum state of a third beam. The first two beams were destroyed in the process, but the third successfully transmitted its properties over a distance of about a yard, Kimble's team reported in the journal *Science.*

Last December a team of physicists in Innsbruck, Austria and a month later another team in Rome said they did a similar thing, with single photons. But Kimble said his team was able to verify what they had done, and also used full light beams as opposed to single photons.

"Ours is an important advance beyond that," he said.

Although the Caltech team worked with light, Kimble thinks teleportation could be applied to solid objects. For instance, the quantum state of a photon could be teleported and applied to a particle, even to an atom.

"Way beyond sex change operations and genetic engineering, the quantum state of one entity could be transported to another entity," Kimble said. "We think we know how to do that."

In other words, an object's individual atoms would not be transported, but transmitting its properties could create a perfect replica.

Could this mean the transporters of the television and movie science-fiction series *Star Trek,* which beam people and objects for huge distances, could one day be a reality?

"I don't think anybody knows the answer," Kimble said. "Let's don't teleport a person—let's teleport the smallest bacterium. How much entanglement would we need to teleport such a thing?"

Would such a teleported bacterium actually be the same bacterium, or just a very good copy?

"Again, no one knows for sure," Kimble said. But his team is working on it.

OK, time travellers—check your time lock, and prepare for takeoff. It looks like we're on our way!

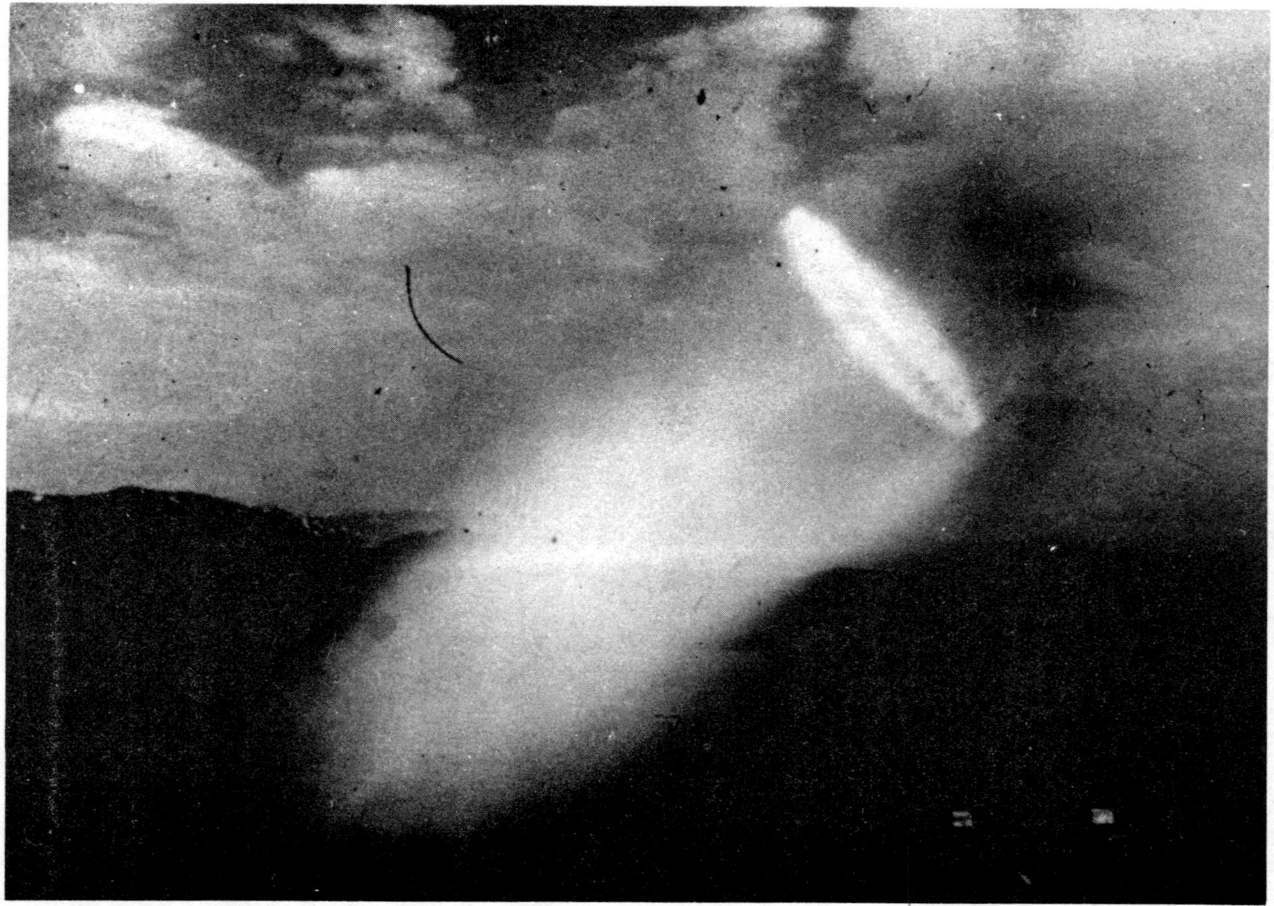

On July 24, 1957, a tourist in Norway was photographing the landscape and this was one of the pictures she took. She was unaware of the UFO on the film until it was developed. Dr. J. Allen Hynek later interviewed her, examined the roll of film, and the photos before and after to be normal. Hynek later wrote, "No explanation for this UFO has been found." Was it a time travel vehicle jumping to hyperspace?

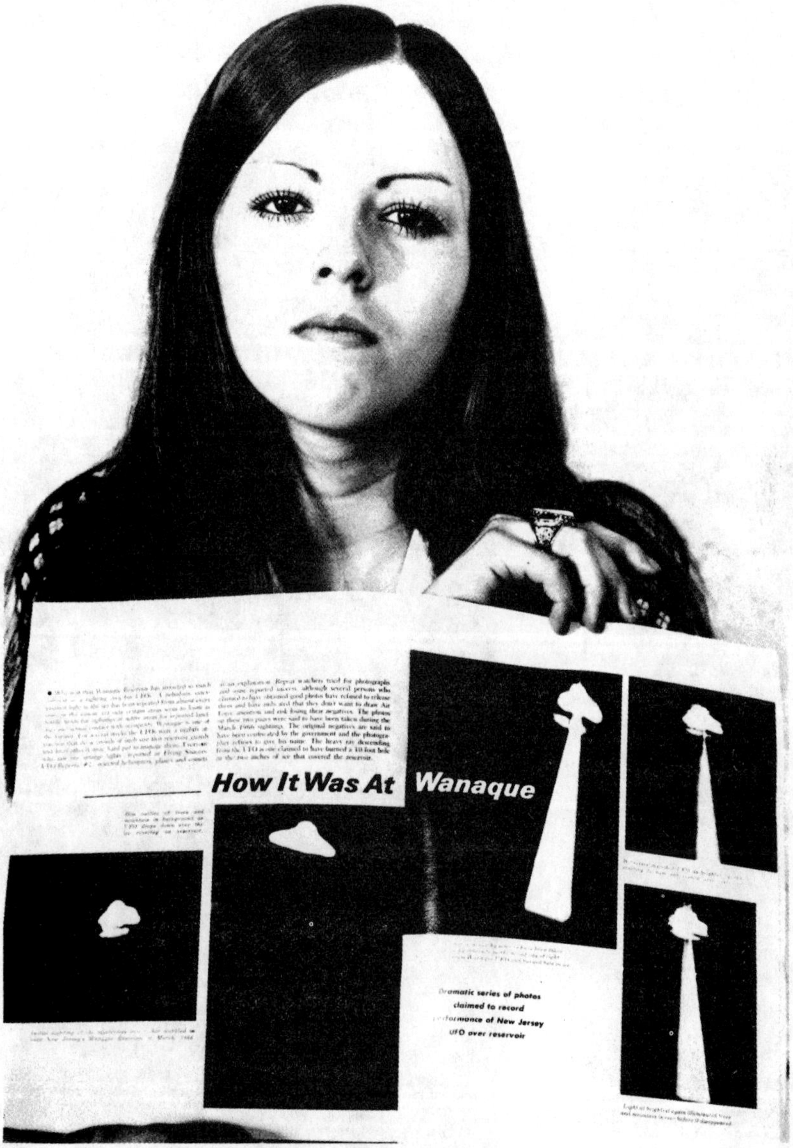

Elain Conroy holds up the True Magazine article about her and the photos taken by her and a police officer of an aircraft in late January of 1966 at the Wanaque Reservoir of New Jersey. Notice the powerful searchlight and apparent vortex of light below the discoid craft.

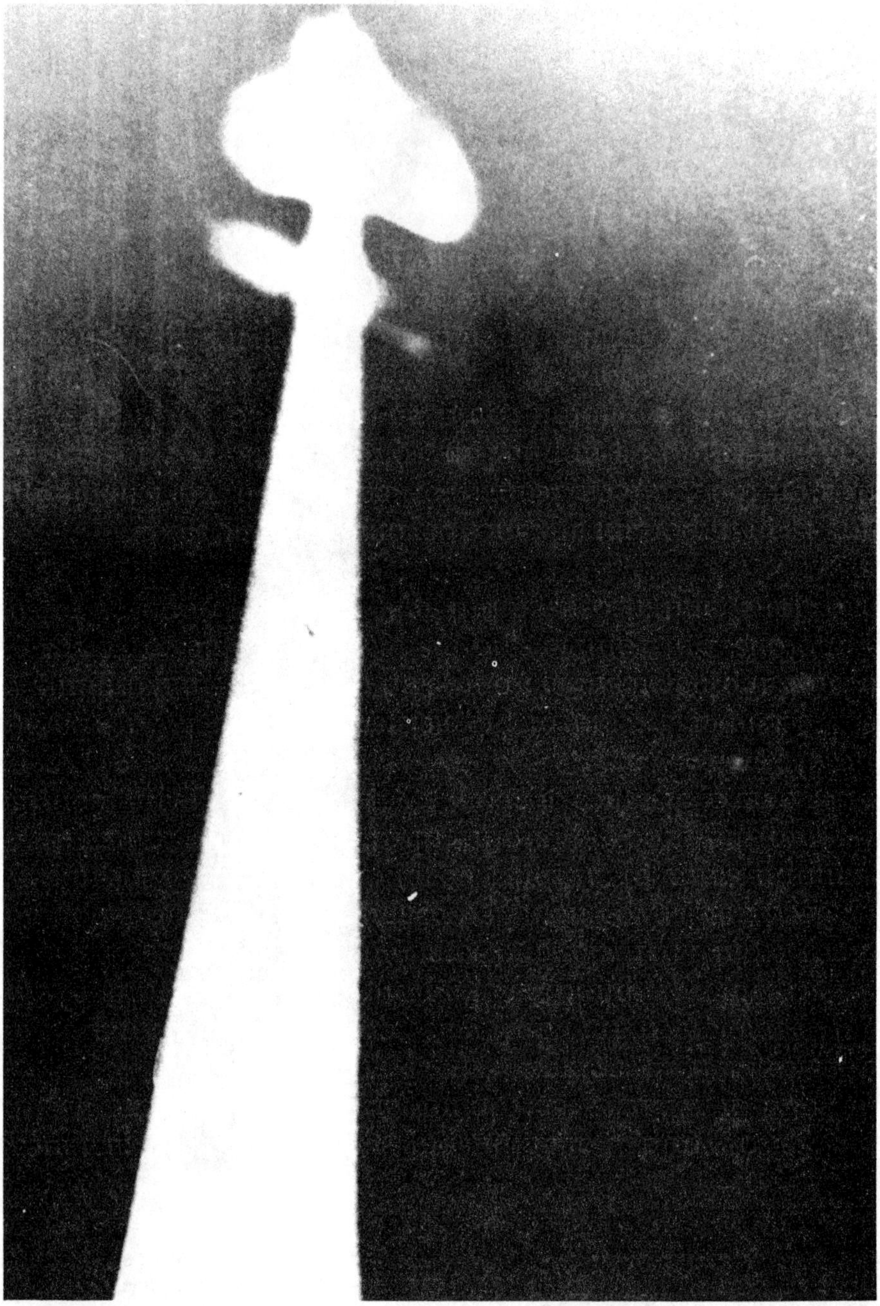

On a late January night in 1966, this photograph of a UFO with a powerful searchlight and an apparent vortex underneath it was taken at the Wanaque Reservoir in New Jersey. Sightings went on for weeks at a time.

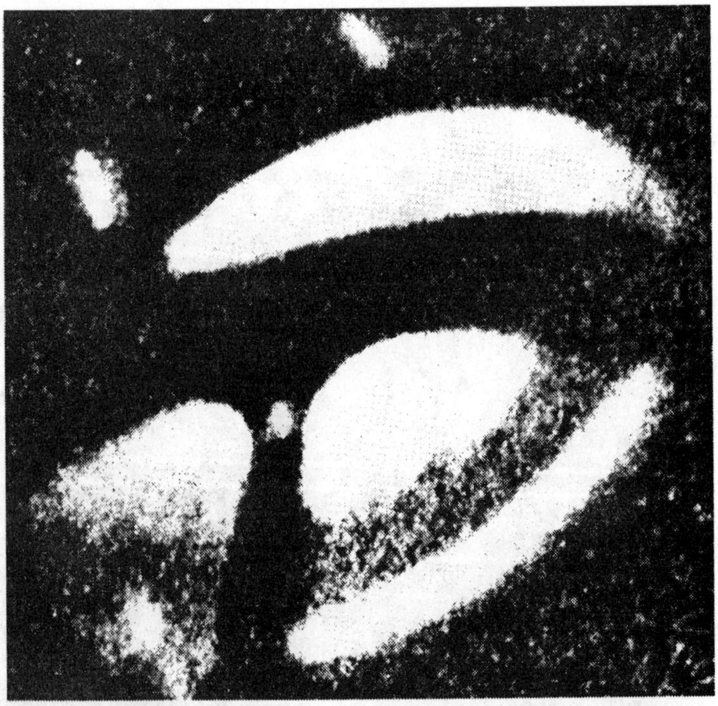

On the night of August 2, 1965, people from the Dakotas to the Mexican border watched brightly colored lights darting across the sky. Sometimes one of the lights would pause for a few seconds. This enabled 14-year-old Alan Smith of Tulas to make a remarkable photo with a small box camera. The negative showed a tiny speck which, when magnified, showed something amazing: a domed UFO divided by dark bands into three segments. Experts said the film was authentic.

This cigar-shaped UFO was photographed in the early 1960s at 12,000 feet by Shinichi Takeda of Fujisaw, Japan. The object appeared to be shadowing the airliner in which Mr. Takeda was travelling, somewhere over Japan. A group of time travel tourists on their way to a dinner reservation?

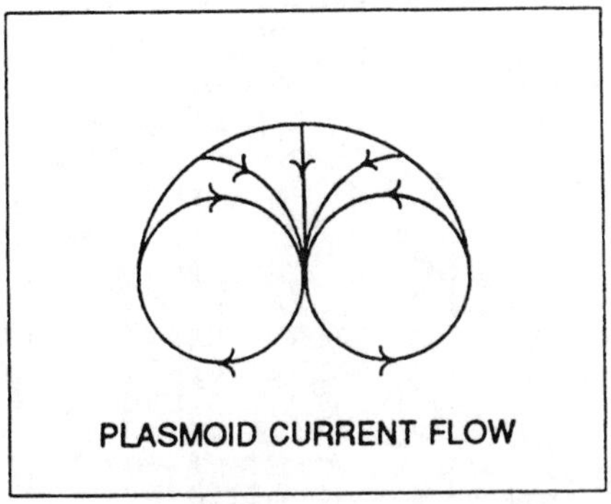

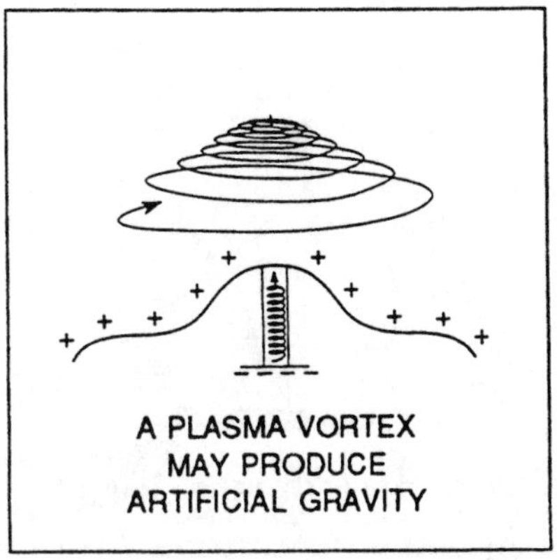

From *Tapping the Zero Point Energy* by Moray B. King.

L. Schroedter, "Vortex Launcher," Private Communication (August 1986).

The following shaped vessel will guide an explosive electrical discharge into a plasma vortex.

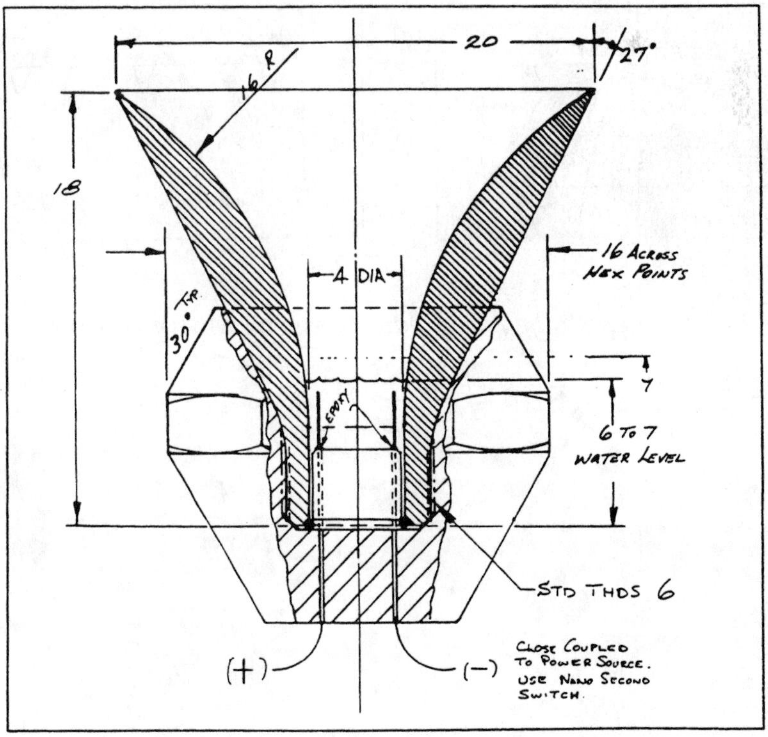

The "Vortex Launcher" designed by L. Schroedter in 1986.

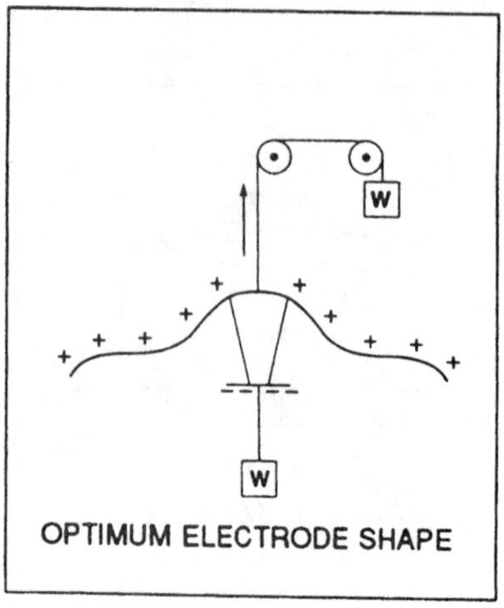

OPTIMUM ELECTRODE SHAPE

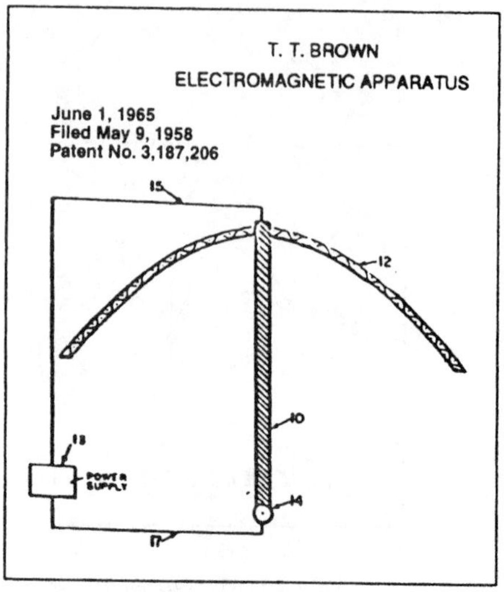

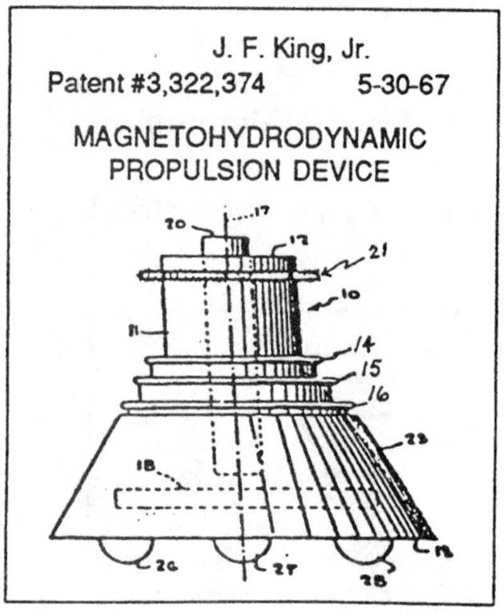

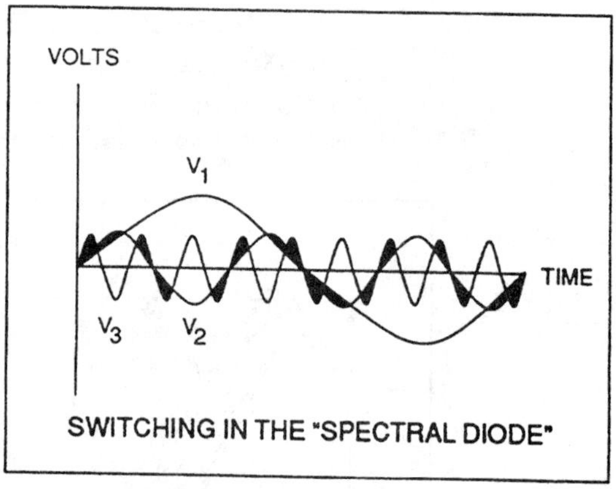

SWITCHING IN THE "SPECTRAL DIODE"

From *Tapping the Zero Point Energy* by Moray B. King.

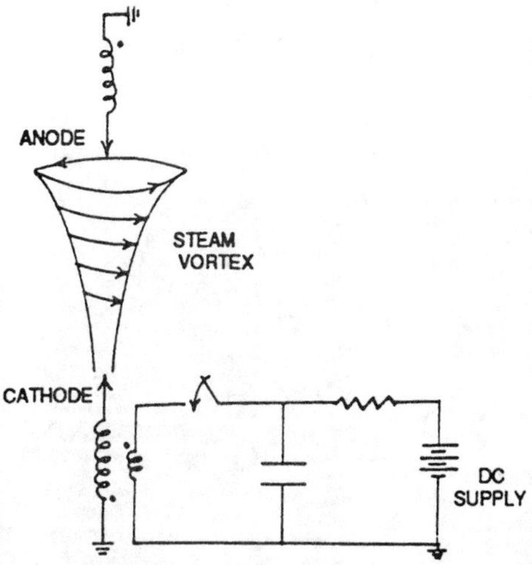

TESLA COIL IN STEAM VORTEX
MAY PRODUCE BALL LIGHTNING

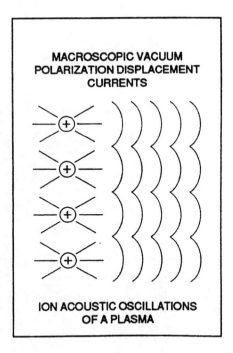

MACROSCOPIC VACUUM
POLARIZATION DISPLACEMENT
CURRENTS

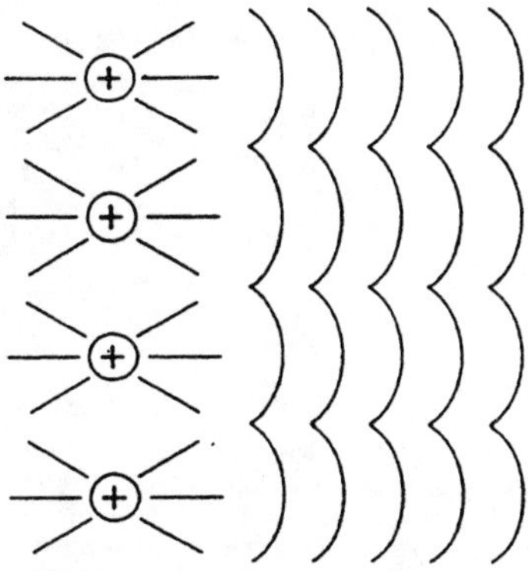

ION ACOUSTIC OSCILLATIONS
OF A PLASMA

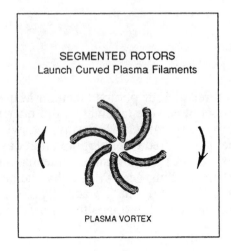

SEGMENTED ROTORS
Launch Curved Plasma Filaments

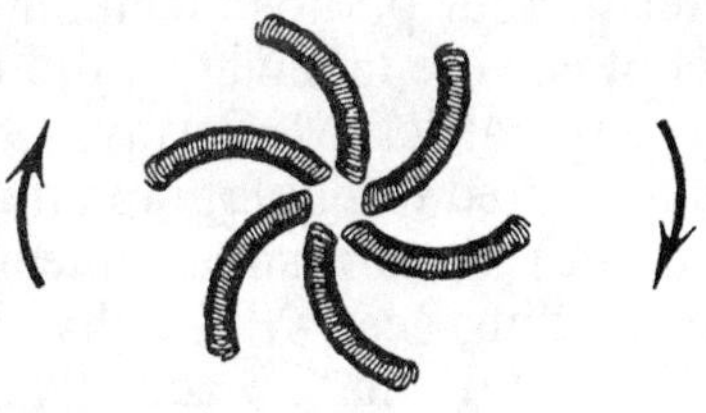

PLASMA VORTEX

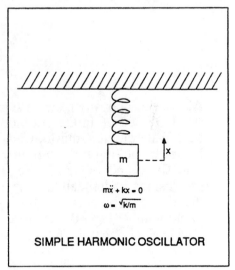

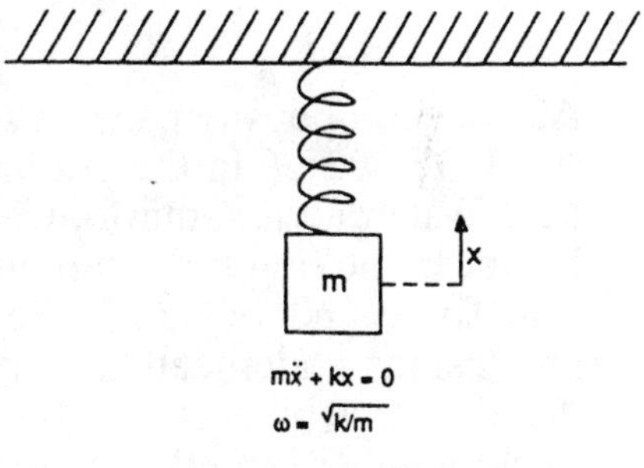

SIMPLE HARMONIC OSCILLATOR

From *Tapping the Zero Point Energy* by Moray B. King.

Above and opposite page: A series of three of four photos taken on March 18, 1975 at 1:30 in the afternoon. The photos were taken at an old quarry near Waterdown, Ontario, Canada, by young Patrick McCarthy, 19, who had been looking for a certain bird he had wanted to photograph that day. The first, and best, of the images (above) shows small antenna-like protrusions underneath the craft, perhaps landing gear. McCarthy claimed that the large object was "flitting around so wildly that he actually missed photo number two. Photos three and four are on the opposite page. A time travel vehicle from the future?

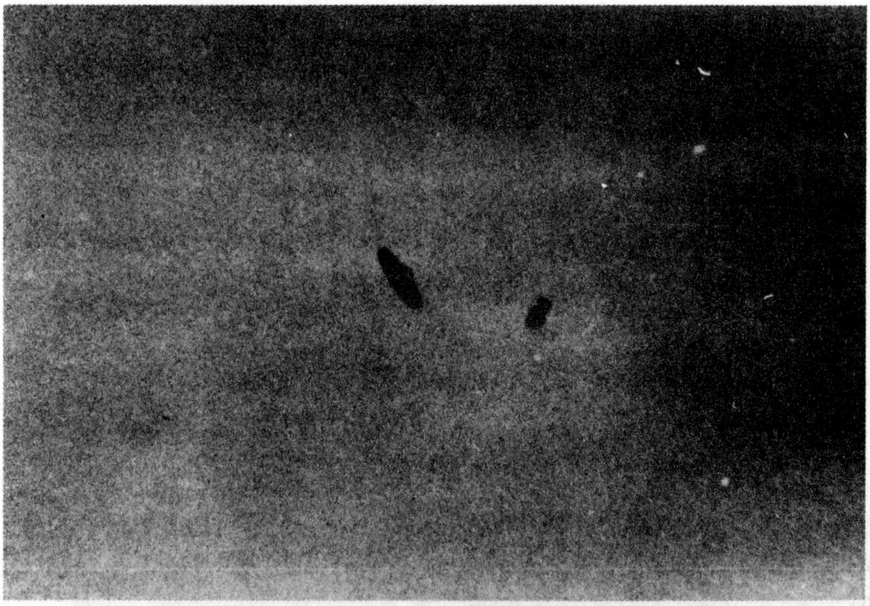

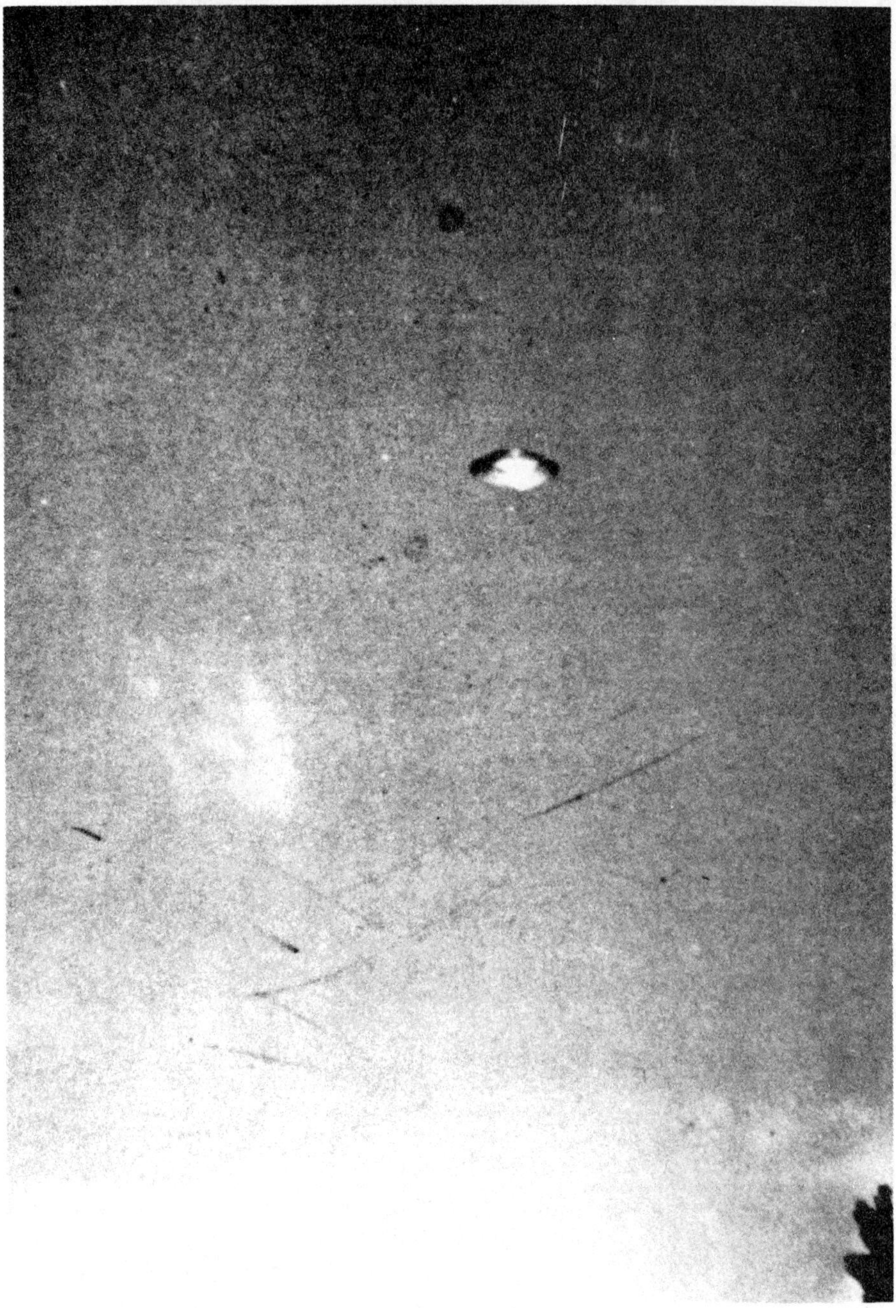

On February 12, 1971, this photo of a flying disc was taken by a French family on North Corsica Island, France. After a slow curving descent the object leveled off and flew toward the nearby Bastia Airfield, a French military base. A time travel anti-gravity device used by the French military?

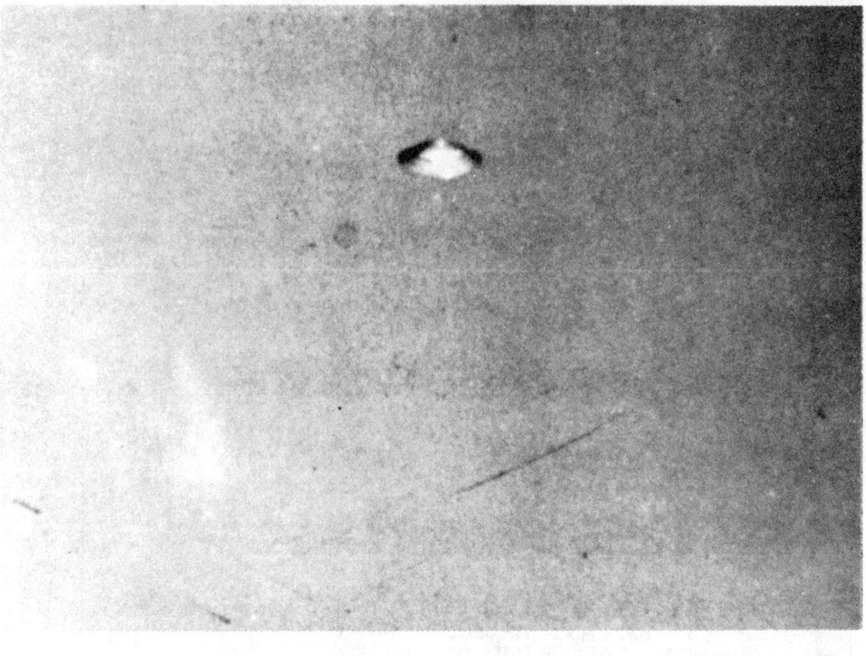

Two more photos taken on February 12, 1971, of a flying disc on North Corsica Island, France. Could this daylight disk have teleported into the area?

A 1938 cartoon "If—You Were Stranded in Time!" by Jack Binder.

A 1938 cartoon "If—You Were Stranded in Time!" by Jack Binder.

The Far Side®

Disaster befalls Professor Schnabel's cleaning lady when she mistakes his time machine for a new dryer.

Calvin and Hobbes **by Bill Watterson**

ONCE AGAIN IT'S TIME to ASK MISTER WHIZBANG!
WHAT'S ALL THIS I HEAR ABOUT THE HUBBLE SPACE TELESCOPE?
SIMPLE!
JIM BORGMAN
NAMED FOR LANKY NEW YORK GIANTS ACE CARL HUBBELL, THIS POWERFUL TELESCOPE WAS RECENTLY SET IN ORBIT BY ASTRONAUTS ABOARD THE SPACE SHUTTLE DIMAGGIO.
WHEN FULLY OPERATIONAL, ITS LARGE OPTICAL MIRROR WILL UNVEIL TO SCIENTISTS A UNIVERSE ALREADY RENDERED PERMANENTLY INCOMPREHENSIBLE BY STEPHEN HAWKING'S BESTSELLER 'TIME, SPACE, BLACK HOLES, DEATH, JUNK BONDS AND METRIC CONVERSIONS'
BY ANALYZING SPACE IMAGES HERETOFORE SEEN ONLY IN 'CALVIN AND HOBBES', MANKIND WILL LOOK 14 BILLION YEARS BACKWARD IN TIME TO OUR OWN ORIGIN AS A LIFELESS CHUNK OF SPACE DEBRIS.
Q: HOW?
A: I ALREADY TOLD YOU: WITH MIRRORS.
WITH THE HUBBLE SPACE TELESCOPE, SCIENTISTS WILL NOW SEE TO THE VERY EDGE OF THE UNIVERSE!
WHAT DO YOU MEAN BY 'THE EDGE OF THE UNIVERSE'?
THAT'S WHERE THE, YOU KNOW, STUFF ENDS.
THEN WHAT'S BEYOND THE EDGE OF THE UNIVERSE?
SAY, ISN'T THAT YOUR MOTHER CALLING?
THE IMPLICATIONS ARE STAGGERING!
GEE, Thanks, Mister WHIZBANG!

the sweet smell of science
BENOIT
DRING DRIIIING
HENRY!
COME AND EAT!
I MADE A REALLY NICE STEW, JUST YOU WAIT AND SEE!
1

CAREFUL! IT'S HOT!
STOP THAT!
HENRY! I INSIST YOU DON'T DO THAT AT THE TABLE!
HENRY!
2

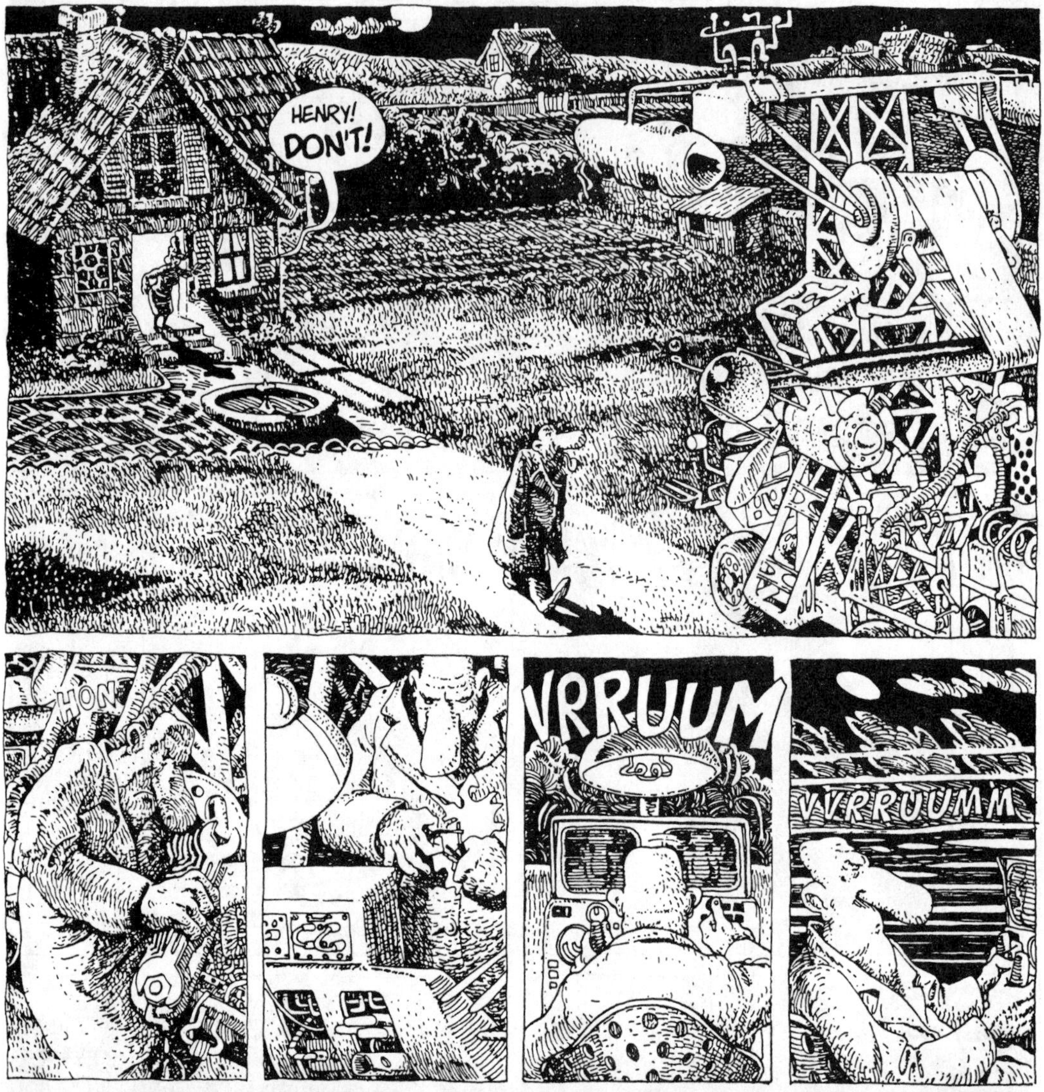
HENRY! DON'T!
VRRUUM
VVRRUUMM

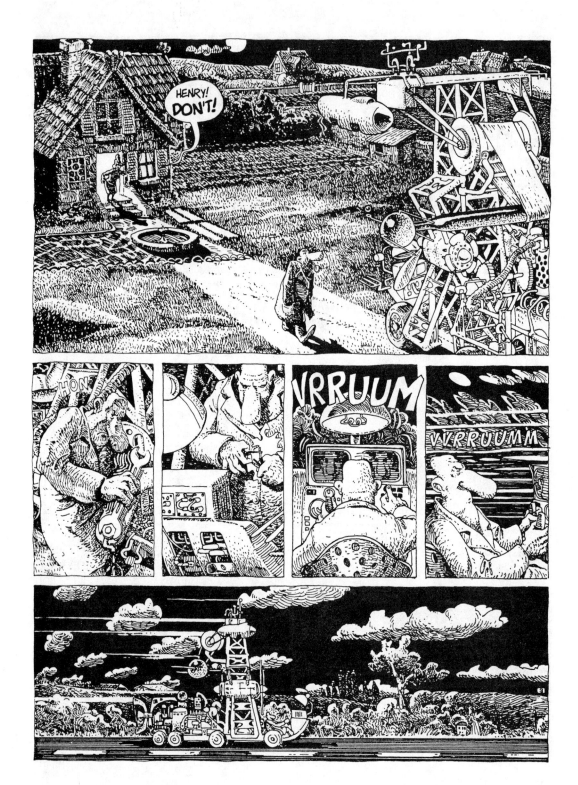

WHAT A RELIEF!
FOR TWO DAYS I'VE BEEN HANGING AROUND IN THE PIT WITH A DRIVE SHAFT ON THE BLINK! OL' BUDDY...
EVER SINCE THEY MADE US THAT NEW SET OF POINTS, YOU CAN'T TELL WHAT'S UP WITH THIS ROAD!
AND MY WIFE IS STILL WAITING FOR ME AT THE CLINIC, POOR THING. SHE'S GOT A BUN IN THE OVEN.
YOUR CRATE SURE IS A HELL OF A MACHINE! CYLINDERS TO THE CAM SHAFT AND ALL THE REST! HO! HO!... COME ON, YOU CAN'T HIDE ANYTHING FROM ME. I'VE GOT AN EXPERT'S EYE.
HEY! IT'S A PLANE, TOO!
FASTEN YOUR SAFETY BELTS!
WHA?!!
4

WAI...
LOUISE!
HOSPITAL
6

THAT... THAT'S MY WIFE! I'M A FATHER! STO... STOP!
STOP THIS FUCKIN' THING!
LOUISE! WAIT FOR ME!
WELL, IF YOU DON'T WANT TO, I'LL TAKE CARE OF IT MYSELF WITH MY SKELETON KEY!
ONE GOOD TURN OF THE SKELETON KEY AND THAT'S IT!
MMM...
6

DID LOUIS SEND YOU? THAT'S A RELIEF!
H
BENG
BELENG
IN THE CAR LOUISE
IN THE LO
BUDUNG
TUDUNG
BREDENG
GRENGELENG
BRE
AND SO: LOUIS ZARZO EXPLORES FUTURE AGES...
WHILE HENRY SCHLITZ-SMITH AND HIS WIFE TAKE IN THE POOR ORPHAN...
JOHN BLAD SINKS A HOLE.
GOOD NEWS IN BOB'S NEWSPAPER:
WOW!
NOT A SINGLE VICE-PRESIDENT WENT NUTS THIS WEEK
7
BENOIT

Bibliography

1. *Time Travel: Myth or Reality,* Richard Heffern, 1977, Pyramid Publications, New York.

2. *Strange Mysteries of Time and Space,* Harold T. Wilkins, 1958, Citadel Books, New York.

3. *Ancient Astronauts: A Time Reversal?*, Robin Collyns, 1976, Sphere Books, London.

4. *Time Travel: Fact, Fiction & Possibility,* Jenny Randles, 1994, Blandford Books, London.

5. *The Ultimate Time Machine*, Joseph McMoneagle, 1998, Hampton Roads Publishing Company, Charlottesville, Virginia.

6. *Time Machines*, Paul J. Nahin, 1993, Springer Verlag, New York.

7. *The Philadelphia Experiment: Project Invisibility*, Charles Berlitz & William L. Moore, 1979, Ballantine Books, New York.

8. *Visitors from Time*, Marc Davenport, 1992, 1994 (second edition), Greenleaf Publications, Tuscaloosa, Alabama.

9. *The Philadelphia Experiment & Other Conspiracies*, Brad Steiger with Alfred Bielek, 1990, Inner Light Publications, New Brunswick, New Jersey.

10. *When Time Breaks Down*, Arthur T. Winfree, 1987, Princeton University Press, Princeton, New Jersey.

11. *Time Warps,* John Gribbin, 1979, Dell Publishing Company, New York.

12. *Exploring the Physics of the Unknown Universe,* Milo Wolff, 1990, Technotron Press, Manhattan Beach, California.

13. *The Cosmic Conspiracy,* Stan Deyo, 1978, West Australian Texas Trading, Kalamunda, Western Australia.

14. *How to Build a Flying Saucer (And Other Proposals in Speculative Engineering)*, T.B. Pawlicki, 1981, Prentice Hall, Inc., Englewood Cliffs, New Jersey.

15. *How You Can Explore the Higher Dimensions of Space and Time (An Introduction to the New Science of Hyperspace for Trekkies of All Ages)*, T. B. Pawlicki, 1984, Prentice Hall, Inc., Englewood Cliffs, New Jersey.

16. *Ether Technology*, Rho Sigma, 1977, Adventures Unlimited Press, Kempton, Illinois.

17. *The Excalibur Briefing*, Thomas E. Bearden, 1980, Walnut Hill Books, San Francisco.

18. *The Philadelphia Experiment Chronicles*, Commander X, 1994, Abelhard Publications, Wilmington, Delaware.

19. *The Aeronauts,* L.T.C. Rolt, 1966, Walker & Co. New York.

20. *The Great Texas Airship Mystery,* Wallace O. Chariton, 1991, Wordware Publishing, Plano, Texas.

21. *Spaceships in Prehistory,* Peter Kolosimo, 1975, University Books, Seacaucus, NJ.

22. *2000 Years of Space Travel,* Russell Freedman, 1963, Collins, London.

23. *Anti-Gravity & the World Grid,* David Hatcher Childress, ed., 1987, Adventures Unlimited Press, Kempton, Illinois.

24. *The UFO Encyclopedia*, Jerome Clark, 1996, Visible Ink, Detroit.

25. *Dimensions,* Jacques Vallee, 1988, Ballantine Books, New York.

26. *Science Frontiers,* William Corliss, 1997, The Sourcebook Project, Glen Arm, Maryland.

27. *Gods of Aquarius (UFO's and the Transformation of Man),* Brad Steiger, 1976, Harcourt Brace Jovanovich, Inc., New York.

29. *Behind the Flying Saucers,* Frank Scully, 1950, Fawcett Books, New York.

30. *Flying Saucers Over Los Angles,* DeWayne B. Johnson and Kenn Thomas, 1950, 2nd edition 1998, Adventures Unlimited Press, Kempton, Illinois.

31. *The Montauk Project*, Preston Nichols and Peter Moon, 1993, Sky Books, Westbury, New York.

32. *Montauk Revisited,* Preston Nichols and Peter Moon, 1994, Sky Books, Westbury, New York.

33. *Pyramids of Montauk*, Preston Nichols and Peter Moon, 1995, Sky Books, Westbury, New York.

34. *Without A Trace,* Charles Berlitz, 1977, Doubleday, New York.

35. *The Case for the UFO,* Morris K. Jessup, 1955, Citadel Press, New York.

36. *The UFO Annual,* edited by Morris K. Jessup, 1956, Citadel Press, New York.

37. *UFO and the Bible,* Morris K. Jessup, 1956, Citadel Press, New York.

38. *Toward a New Electromagnetics, Parts 1,2,3 & 4,* 1983, Thomas E. Bearden, Tesla Book Company, Millbrae, California.

39. *Grolier Multimedia Encyclopedia*, 1997, Grolier Interactive, Inc., Danbury, Connecticut.

40. *Time Travelers from Our Future,* Dr. Bruce Goldberg, 1998, Llewllyn Publications, St. Paul, Minnesota.

41. *M.K. Jessup, the Allende Letters, and Gravity*, Riley Crabb, 1962, Borderland Sciences Research Foundation, Vista, California.

42. *One Hundred Thousand Years of Man's Unknown History,* Robert Charroux, 1963, Berkley Publishing Corporation, New York.

43. *We Are Not the First*, Andrew Tomas, 1971, Bantam Books, Inc., New York.

44. *Solutions to Tesla's Secrets and the Soviet Tesla Weapons,* Thomas E. Bearden, 1983, Tesla Book Company, Millbrae, California.

45. *A Dual Ether Universe,* Leonid Sokolow, 1977, Exposition Press, Inc., Hicksville, New York.

46. *Stalking the Wild Pendulum,* Itzhak Bentov, 1977, E. P. Dutton, New York.

47. *Alternative (003),* Leslie Watkins, 1977, 1978, Avon Books, New York. (originally a television show in Britain.)

48. *The Energy Grid,* Bruce L. Cathie, 1995, Adventures Unlimited Press, Kempton, Illinois.

49. *The Bridge to Infinity,* Bruce L. Cathie, 1983, Adventures Unlimited Press, Kempton, Illinois.

50. *The Harmonic Conquest of Space,* Bruce L. Cathie, 1998, Adventures Unlimited Press, Kempton, Illinois.

51. *Flying Saucers—Serious Business,* Frank Edwards, 1966, Bantam Books, Inc., New York.

52. *Somebody Else is on the Moon,* George Leonard, 1976, 1977, Pocket Books, a Simon & Schuster Division, New York.

53. *The Roswell Incident,* Charles Berlitz & William L. Moore, 1980, Grosset & Dunlap (A Filmways Company), New York.

54. *Our Mysterious Spaceship Moon,* Don Wilson, 1975, Dell Publishing Co., Inc. New York.

55. *Secrets of Our Spaceship Moon,* Don Wilson, 1979, Dell Publishing Co., Inc., New York.

56. *Messengers of Deception,* Jacques Vallee, 1979, 1980, Bantam Books, Inc., New York.

57. *Mysteries of Time and Space,* Brad Steiger, 1974, Dell Publishing Co., Inc. New York.

58. *Challenge to Science: The UFO Enigma,* Jacques & Janine Vallee, 1966, Ballantine Books, a Division of Random House, Inc., New York.

59. *Profiles of the Future,* Arthur C. Clarke, 1958, 59, 60, 62, Harper & Row, Inc., New York.

60. *Mysticism and the New Physics,* Michael Talbot, 1980, Bantam Books, Inc., New York.

61. *UFO's Past, Present & Future,* Robert Emenegger, 1974, Ballantine Books, a Division of Random House, Inc., New York.

62. *Beyond Earth: Man's Contact with UFO's,* Ralph Blum with Judy Blum, 1974, Bantam Books, Inc., New York.

63. *The House of Lords UFO Debate,* Lord Clancarty, 1979, Open Head Press, London, in association with Pentacle Books, Bristol, Great Britain.

64. *New Horizons in Electric, Magnetic and Gravitational Field Theory,* W. J. Hooper, PhD, 1974, Electrodynamic Gravity, Inc., Cuyahoga Falls, Ohio.

65. *Future Physics and Anti-Gravity*, William F. Hassel, PhD., MUFON Symposium Proceedings of 16,17 July 1977, Woodland Hills, California.

66. "Should the Laws of Gravitation Be Reconsidered?, " Maurice F. G. Allasi, *Aero/Space Engineering*, Sept. 1959, pp. 46-52; Oct. 1959, pp. 51-55; Nov. 1959, p. 55.

67. *The Sea of Energy in Which the Earth Floats,* Henry T. Moray, 1960, Utah.

68. "Particles That Travel Faster Than Light, " R. G. Newton, *Science*, Vol. 167, No. 3925, Mar. 20, 1970, pp. 1569-1574.

69. "UFO's: Theories in Time Travel, Dimensions, and Anti-Universes, " John M. Prytz, *Flying Saucers*, Dec. 1969, pp. 4-7.

70. *The Principles of Ultra-Relativity*, Shinichi Seike, 1970, National Space Research Consortium, Uwajima City, Japan.

72. *Space, Time and Gravitation,* W. Kopczynski and A. Trautman, 1992, MacMillan, New York.

73. *The Philosophy of Time,* Robin Le Poidevin and Murray Macbeath, eds., 1993, MacMillan, New York.

THE ADVENTURES UNLIMITED CATALOG

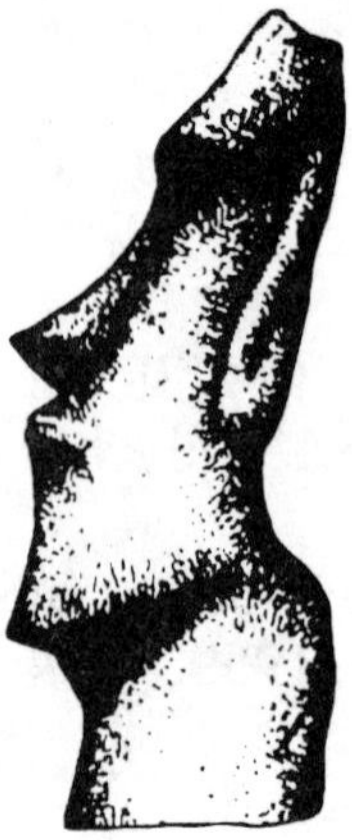

ANTI-GRAVITY

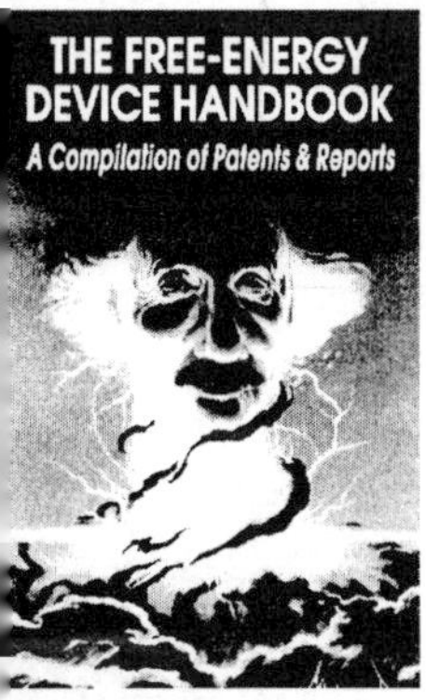

THE FREE-ENERGY DEVICE HANDBOOK

A Compilation of Patents and Reports

by David Hatcher Childress

A large-format compilation of various patents, papers, descriptions and diagrams concerning free-energy devices and systems. *The Free-Energy Device Handbook* is a visual tool for experimenters and researchers into magnetic motors and other "over-unity" devices. With chapters on the Adams Motor, the Hans Coler Generator, cold fusion, superconductors, "N" machines, space-energy generators, Nikola Tesla, T. Townsend Brown, and the latest in free-energy devices. Packed with photos, technical diagrams, patents and fascinating information, this book belongs on every science shelf. With energy and profit being a major political reason for fighting various wars, free-energy devices, if ever allowed to be mass distributed to consumers, could change the world! Get your copy now before the Department of Energy bans this book!

292 PAGES. 8X10 PAPERBACK. ILLUSTRATED. BIBLIOGRAPHY. $16.95. CODE: FEH

THE ANTI-GRAVITY HANDBOOK

edited by David Hatcher Childress, with Nikola Tesla, T.B. Paulicki, Bruce Cathie, Albert Einstein and others

The new expanded compilation of material on Anti-Gravity, Free Energy, Flying Saucer Propulsion, UFOs, Suppressed Technology, NASA Cover-ups and more. Highly illustrated with patents, technical illustrations and photos. This revised and expanded edition has more material, including photos of Area 51, Nevada, the government's secret testing facility. This classic on weird science is back in a 90s format!

- **How to build a flying saucer.**
- **Arthur C. Clarke on Anti-Gravity.**
- **Crystals and their role in levitation.**
- **Secret government research and development.**
- **Nikola Tesla on how anti-gravity airships could draw power from the atmosphere.**
- **Bruce Cathie's Anti-Gravity Equation.**
- **NASA, the Moon and Anti-Gravity.**

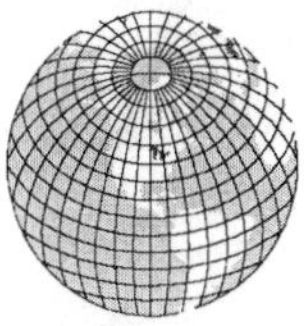

230 PAGES. 7X10 PAPERBACK. BIBLIOGRAPHY/INDEX/APPENDIX. HIGHLY ILLUSTRATED. $14.95. CODE: AGH

ANTI–GRAVITY & THE WORLD GRID

Is the earth surrounded by an intricate electromagnetic grid network offering free energy? This compilation of material on ley lines and world power points contains chapters on the geography, mathematics, and light harmonics of the earth grid. Learn the purpose of ley lines and ancient megalithic structures located on the grid. Discover how the grid made the Philadelphia Experiment possible. Explore the Coral Castle and many other mysteries, including acoustic levitation, Tesla Shields and scalar wave weaponry. Browse through the section on anti-gravity patents, and research resources.

274 PAGES. 7X10 PAPERBACK. ILLUSTRATED. $14.95. CODE: AGW

ANTI–GRAVITY & THE UNIFIED FIELD

edited by David Hatcher Childress

Is Einstein's Unified Field Theory the answer to all of our energy problems? Explored in this compilation of material is how gravity, electricity and magnetism manifest from a unified field around us. Why artificial gravity is possible; secrets of UFO propulsion; free energy; Nikola Tesla and anti-gravity airships of the 20s and 30s; flying saucers as superconducting whirls of plasma; anti-mass generators; vortex propulsion; suppressed technology; government cover-ups; gravitational pulse drive; spacecraft & more.

240 PAGES. 7X10 PAPERBACK. ILLUSTRATED. $14.95. CODE: AGU

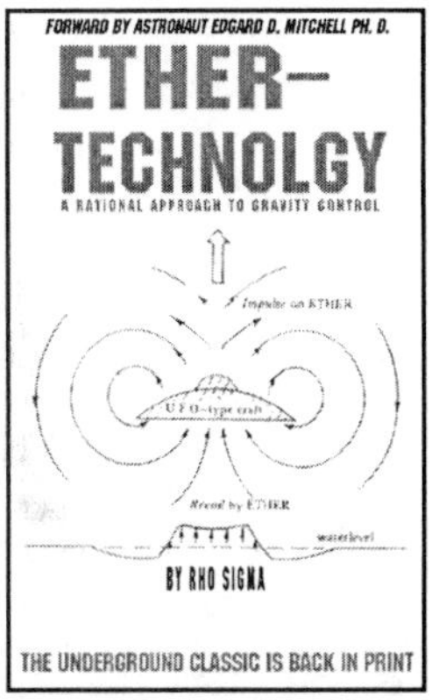

FORWARD BY ASTRONAUT EDGARD D. MITCHELL PH. D.
ETHER-TECHNOLGY
A RATIONAL APPROACH TO GRAVITY CONTROL
BY RHO SIGMA
THE UNDERGROUND CLASSIC IS BACK IN PRINT

ETHER TECHNOLOGY

A Rational Approach to Gravity Control

by Rho Sigma

This classic book on anti-gravity and free energy is back in print and back in stock. Written by a well-known American scientist under the pseudonym of "Rho Sigma," this book delves into international efforts at gravity control and discoid craft propulsion. Before the Quantum Field, there was "Ether." This small, but informative book has chapters on John Searle and "Searle discs;" T. Townsend Brown and his work on anti-gravity and ether-vortex turbines. Includes a forward by former NASA astronaut Edgar Mitchell.

108 PAGES. 6X9 PAPERBACK. ILLUSTRATED. $12.95. CODE: ETT

ANTI-GRAVITY

COSMIC MATRIX

Piece for a Jig-Saw, Part Two

by Leonard G. Cramp

Leonard G. Cramp, a British aerospace engineer, wrote his first book *Space Gravity and the Flying Saucer* in 1954. Cosmi Matrix is the long-awaited sequel to his 1966 book *UFOs & Anti-Gravity: Piece for a Jig-Saw.* Cramp has had a long histo of examining UFO phenomena and has concluded that UFOs use the highest possible aeronautic science to move in the wa they do. Cramp examines anti-gravity effects and theorizes that this super-science used by the craft—described in detail in th book—can lift mankind into a new level of technology, transportation and understanding of the universe. The book takes close look at gravity control, time travel, and the interlocking web of energy between all planets in our solar system wi Leonard's unique technical diagrams. A fantastic voyage into the present and future!

364 PAGES. 6X9 PAPERBACK. ILLUSTRATED. BIBLIOGRAPHY. $16.00. CODE: CMX

UFOS AND ANTI-GRAVITY

Piece For A Jig-Saw

by Leonard G. Cramp

Leonard G. Cramp's 1966 classic book on flying saucer propulsion and suppressed technology is a highly technical look at the UFO phenomena by a trained scientist. Cramp first introduces the idea of 'anti-gravity' and introduces us to the various theories of gravitation. He then examines the technology necessary to build a flying saucer and examines in great detail the technical aspects of such a craft. Cramp's book is a wealth of material and diagrams on flying saucers, anti-gravity, suppressed technology, G-fields and UFOs. Chapters include Crossroads of Aerodymanics, Aerodynamic Saucers, Limitations of Rocketry, Gravitation and the Ether, Gravitational Spaceships, G-Field Lift Effects, The Bi-Field Theory, VTOL and Hovercraft, Analysis of UFO photos, more.

388 PAGES. 6X9 PAPERBACK. ILLUSTRATED. $16.95. CODE: UAG

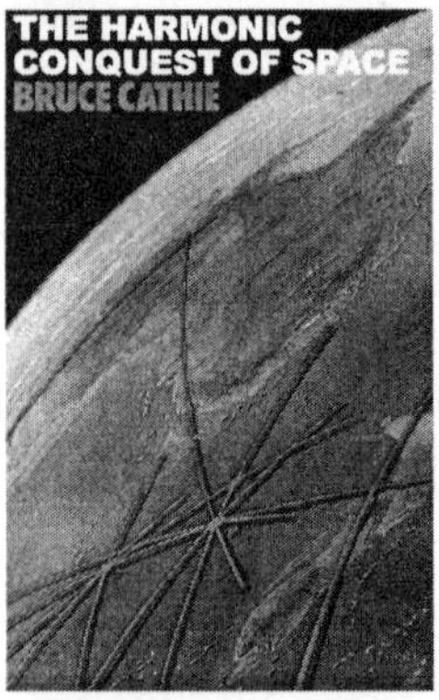

THE HARMONIC CONQUEST OF SPACE

by Captain Bruce Cathie

Chapters include: Mathematics of the World Grid; the Harmonics of Hiroshima and Nagasaki; Harmonic Transmission and Receiving; the Link Between Human Brain Waves; the Cavity Resonance between the Earth; the Ionosphere and Gravity; Edgar Cayce—the Harmonics of the Subconscious; Stonehenge; the Harmonics of the Moon; the Pyramids of Mars; Nikola Tesla's Electric Car; the Robert Adam Pulsed Electric Motor Generator; Harmonic Clues to the Unified Field; and more. Also included are tables showing th harmonic relations between the earth's magnetic field, the speed of light, and anti-gravity/gravity acceleration at differe points on the earth's surface. New chapters in this edition on the giant stone spheres of Costa Rica, Atomic Tests and Volcan Activity, and a chapter on Ayers Rock analysed with Stone Mountain, Georgia.

248 PAGES. 6X9. PAPERBACK. ILLUSTRATED. BIBLIOGRAPHY. $16.95. CODE: HCS

THE ENERGY GRID

Harmonic 695, The Pulse of the Universe

by Captain Bruce Cathie.

This is the breakthrough book that explores the incredible potential of the Energy Grid and the Earth's Unified Field all arou us. Cathie's first book, *Harmonic 33*, was published in 1968 when he was a commercial pilot in New Zealand. Since th Captain Bruce Cathie has been the premier investigator into the amazing potential of the infinite energy that surrounds c planet every microsecond. Cathie investigates the Harmonics of Light and how the Energy Grid is created. In this amazi book are chapters on UFO Propulsion, Nikola Tesla, Unified Equations, the Mysterious Aerials, Pythagoras & the Grid, Nucl Detonation and the Grid, Maps of the Ancients, an Australian Stonehenge examined, more.

255 PAGES. 6X9 TRADEPAPER. ILLUSTRATED. $15.95. CODE: TEG

THE BRIDGE TO INFINITY

Harmonic 371244

by Captain Bruce Cathie

Cathie has popularized the concept that the earth is crisscrossed by an electromagnetic grid system that can be used for anti-gravity, free energy, levitation and more. The book includes a new analysis of the harmonic nature of reality, acoustic levitation, pyramid power, harmonic receiver towers and UFO propulsion. It concludes that today's scientists have at their command a fantastic store of knowledge with which to advance the welfare of the human race.

204 PAGES. 6X9 TRADEPAPER. ILLUSTRATED. $14.95. CODE: BTF

MAN-MADE UFOS 1944—1994

Fifty Years of Suppression

by Renato Vesco & David Hatcher Childress

A comprehensive look at the early "flying saucer" technology of Nazi Germany and the genesis of man-made UFOs. This book takes us from the work of captured German scientists to escaped battalions of Germans, secr communities in South America and Antarctica to todays state-of-the-art "Dreamland" flying machines. Heavily illustrate this astonishing book blows the lid off the "government UFO conspiracy" and explains with technical diagrams the techno ogy involved. Examined in detail are secret underground airfields and factories; German secret weapons; "suction" aircra the origin of NASA; gyroscopic stabilizers and engines; the secret Marconi aircraft factory in South America; and mor Introduction by W.A. Harbinson, author of the Dell novels *GENESIS* and *REVELATION*.

318 PAGES. 6X9 PAPERBACK. ILLUSTRATED. INDEX & FOOTNOTES. $18.95. CODE: MMU

TESLA TECHNOLOGY

THE FANTASTIC INVENTIONS OF NIKOLA TESLA

Nikola Tesla with additional material by David Hatcher Childress

This book is a readable compendium of patents, diagrams, photos and explanations of the many incredible inventions of the originator of the modern era of electrification. In Tesla's own words are such topics as wireless transmission of power, death rays, and radio-controlled airships. In addition, rare material on German bases in Antarctica and South America, and a secret city built at a remote jungle site in South America by one of Tesla's students, Guglielmo Marconi. Marconi's secret group claims to have built flying saucers in the 1940s and to have gone to Mars in the early 1950s! Incredible photos of these Tesla craft are included. The Ancient Atlantean system of broadcasting energy through a grid system of obelisks and pyramids is discussed, and a fascinating concept comes out of one chapter: that Egyptian engineers had to wear protective metal head-shields while in these power plants, hence the Egyptian Pharoah's head covering as well as the Face on Mars!

•His plan to transmit free electricity into the atmosphere. •How electrical devices would work using only small antennas mounted on them. •Why unlimited power could be utilized anywhere on earth. •How radio and radar technology can be used as death-ray weapons in Star Wars. •Includes an appendix of Supreme Court documents on dismantling his free energy towers. •Tesla's Death Rays, Ozone generators, and more…

342 PAGES. 6X9 PAPERBACK. ILLUSTRATED. BIBLIOGRAPHY AND APPENDIX. $16.95. CODE: FINT

THE TESLA PAPERS

Nikola Tesla on Free Energy & Wireless Transmission of Power

by Nikola Tesla, edited by David Hatcher Childress

In the tradition of *The Fantastic Inventions of Nikola Tesla, The Anti-Gravity Handbook* and *The Free-Energy Device Handbook,* science and UFO author David Hatcher Childress takes us into the incredible world of Nikola Tesla and his amazing inventions. Tesla's rare article "The Problem of Increasing Human Energy with Special Reference to the Harnessing of the Sun's Energy" is included. This lengthy article was originally published in the June 1900 issue of *The Century Illustrated Monthly Magazine* and it was the outline for Tesla's master blueprint for the world. Tesla's fantastic vision of the future, including wireless power, anti-gravity, free energy and highly advanced solar power.

Also included are some of the papers, patents and material collected on Tesla at the Colorado Springs Tesla Symposiums, including papers on:

•The Secret History of Wireless Transmission •Tesla and the Magnifying Transmitter
•Design and Construction of a half-wave Tesla Coil •Electrostatics: A Key to Free Energy
•Progress in Zero-Point Energy Research •Electromagnetic Energy from Antennas to Atoms
•Tesla's Particle Beam Technology •Fundamental Excitatory Modes of the Earth-Ionosphere Cavity

325 PAGES. 8X10 PAPERBACK. ILLUSTRATED. $16.95. CODE: TTP

LOST SCIENCE

by Gerry Vassilatos

Secrets of Cold War Technology author Vassilatos on the remarkable lives, astounding discoveries, and incredible inventions of such famous people as Nikola Tesla, Dr. Royal Rife, T.T. Brown, and T. Henry Moray. Read about the aura research of Baron Karl von Reichenbach, the wireless technology of Antonio Meucci, the controlled fusion devices of Philo Farnsworth, the earth battery of Nathan Stubblefield, and more. What were the twisted intrigues which surrounded the often deliberate attempts to stop this technology? Vassilatos claims that we are living hundreds of years behind our intended level of technology and we must recapture this "lost science."

304 PAGES. 6X9 PAPERBACK. ILLUSTRATED. BIBLIOGRAPHY. $16.95. CODE: LOS

SECRETS OF COLD WAR TECHNOLOGY

Project HAARP and Beyond

by Gerry Vassilatos

Vassilatos reveals that "Death Ray" technology has been secretly researched and developed since the turn of the century. Included are chapters on such inventors and their devices as H.C. Vion, the developer of auroral energy receivers; Dr. Selim Lemstrom's pre-Tesla experiments; the early beam weapons of Grindell-Mathews, Ulivi, Turpain and others; John Hettenger and his early beam power systems. Learn about Project Argus, Project Teak and Project Orange; EMP experiments in the 60s; why the Air Force directed the construction of a huge Ionospheric "backscatter" telemetry system across the Pacific just after WWII; why Raytheon has collected every patent relevant to HAARP over the past few years; more.

250 PAGES. 6X9 PAPERBACK. ILLUSTRATED. $15.95. CODE: SCWT

HAARP

The Ultimate Weapon of the Conspiracy

by Jerry Smith

The HAARP project in Alaska is one of the most controversial projects ever undertaken by the U.S. Government. Jerry Smith gives us the history of the HAARP project and explains how works, in technically correct yet easy to understand language. At best, HAARP is science out-of-control; at worst, HAARP could be the most dangerous device ever created, a futuristic technology that is everything from super-beam weapon to world-wide mind control device. Topics include Over-the-Horizon Radar and HAARP, Mind Control, ELF and HAARP, The Telsa Connection, The Russian Woodpecker, GWEN & HAARP, Earth Penetrating Tomography, Weather Modification, Secret Science of the Conspiracy, more. Includes the complete 1987 Eastlund patent for his pulsed super-weapon that he claims was stolen by the HAARP Project.

256 PAGES. 6X9 PAPERBACK. ILLUSTRATED. $14.95. CODE: HARP

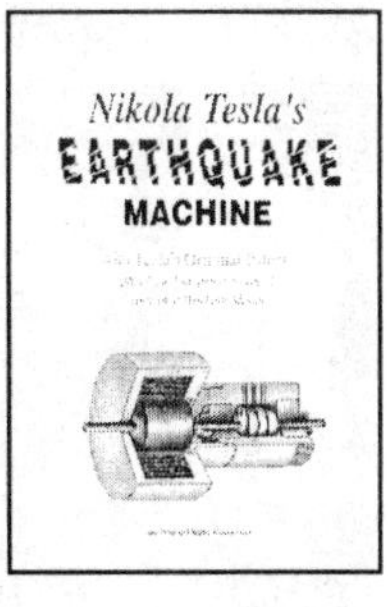

NIKOLA TESLA'S EARTHQUAKE MACHINE

with Tesla's Original Patents

by Dale Pond and Walter Baumgartner

Now, for the first time, the secrets of Nikola Tesla's Earthquake Machine are available. Although this book discusses in detail Nikola Tesla's 1894 "Earthquake Oscillator," it is also about the new technology of sonic vibrations which produce a resonance effect that can be used to cause earthquakes. Discussed are Tesla Oscillators, Vibration Physics, Amplitude Modulated Additive Synthesis, Tele-Geo-dynamics, Solar Heat Pump Apparatus, Vortex Tube Coolers, the Serogodsky Motor, more. Plenty of technical diagrams. Be the first on your block to have a Tesla Earthquake Machine!

175 PAGES. 9X11 PAPERBACK. ILLUSTRATED. BIBLIOGRAPHY & INDEX. $16.95. CODE: TEM

STRANGE SCIENCE

UNDERGROUND BASES & TUNNELS

What is the Government Trying to Hide?

by Richard Sauder, Ph.D.

Working from government documents and corporate records, Sauder has compiled an impressive book that digs below the surface of the military's super-secret underground! Go behind the scenes into little-known corners of the public record and discover how corporate America has worked hand-in-glove with the Pentagon for decades, dreaming about, planning, and actually constructing, secret underground bases. This book includes chapters on the locations of the bases, the tunneling technology, various military designs for underground bases, nuclear testing & underground bases, abductions, needles & implants, military involvement in "alien" cattle mutilations, more. 50 page photo & map insert.

201 PAGES. 6X9 PAPERBACK. ILLUSTRATED. $15.95. CODE: UGB

UNDERWATER & UNDERGROUND BASES

Surprising Facts the Government Does Not Want You to Know

by Richard Sauder

Dr. Richard Sauder's brand new book *Underwater and Underground Bases* is an explosive, eye-opening sequel to his best-selling, *Undergrou Bases and Tunnels: What is the Government Trying to Hide?* Dr. Sauder lays out the amazing evidence and government paper trail for the constru tion of huge, manned bases offsore, in mid-ocean, and deep beneath the sea floor! Bases big enough to secretly dock submarines! Official Uni States Navy documents, and other hard evidence, raise many questions about what really lies 20,000 leagues beneath the sea. Many UFOs have be seen coming and going from the world's oceans, seas and lakes, implying the existence of secret underwater bases. Hold on to your hats: Jules Ve may not have been so far from the truth, after all! Dr. Sauder also adds to his incredible database of underground bases onshore. New, breakthrou material reveals the existence of additional clandestine underground facilities as well as the surprising location of one of the CIA's own undergrou bases. Plus, new information on tunneling and cutting-edge, high speed rail magnetic-levitation (MagLev) technology. There are many rumors secret, underground tunnels with MagLev trains hurtling through them. Is there truth behind the rumors? *Underwater and Underground Ba* carefully examines the evidence and comes to a thought provoking conclusion!

264 PAGES. 6X9 PAPERBACK. ILLUSTRATED. BIBLIOGRAPHY. INDEX. $16.95. CODE: UUB

KUNDALINI TALES

by Richard Sauder, Ph.D.

Underground Bases and Tunnels author Richard Sauder's second book on his personal experiences and provocative research into sp taneous spiritual awakening, out-of-body journeys, encounters with secretive governmental powers, daylight sightings of UFOs, a more. Sauder continues his studies of underground bases with new information on the occult underpinnings of the U.S. space progra The book also contains a breakthrough section that examines actual U.S. patents for devices that manipulate minds and thoughts fro remote distance. Included are chapters on the secret space program and a 130-page appendix of patents and schematic diagrams of sec technology and mind control devices.

296 PAGES. 7X10 PAPERBACK. ILLUSTRATED. BIBLIOGRAPHY. $14.95. CODE: KTAL

THE TIME TRAVEL HANDBOOK

A Manual of Practical Teleportation & Time Travel

edited by David Hatcher Childress

In the tradition of *The Anti-Gravity Handbook* and *The Free-Energy Device Handbook,* science and UFO author David Hatcher Childre takes us into the weird world of time travel and teleportation. Not just a whacked-out look at science fiction, this book is an authoritati chronicling of real-life time travel experiments, teleportation devices and more. *The Time Travel Handbook* takes the reader beyond t government experiments and deep into the uncharted territory of early time travellers such as Nikola Tesla and Guglielmo Marconi and the alleged time travel experiments, as well as the Wilson Brothers of EMI and their connection to the Philadelphia Experiment—the U.S. Navy forays into invisibility, time travel, and teleportation. Childress looks into the claims of time travelling individuals, and investigates the unusu claim that the pyramids on Mars were built in the future and sent back in time. A highly visual, large format book, with patents, photos ar schematics. Be the first on your block to build your own time travel device!

316 PAGES. 7X10 PAPERBACK. ILLUSTRATED. $16.95. CODE: TTH

MAPS OF THE ANCIENT SEA KINGS

Evidence of Advanced Civilization in the Ice Age

by Charles H. Hapgood

Charles Hapgood's classic 1966 book on ancient maps produces concrete evidence of an advanced world-wide civilization existing many thousands of years before ancient Egypt. He has found the evidence in the Piri Reis Map that shows Antarctica, the Hadji Ahmed map, the Oronteus Finaeus and other amazing maps. Hapgood concluded that these maps were made from more ancient maps from the various ancient archives around the world, now lost. Not only were these unknown people more advanced in mapmaking than any people prior to the 18th century, it appears they mapped all the continents. The Americas were mapped thousands of years before Columbus. Antarctica was mapped when its coasts were free of ice.

316 PAGES. 7X10 PAPERBACK. ILLUSTRATED. BIBLIOGRAPHY & INDEX. $19.95. CODE: MASK

PATH OF THE POLE

Cataclysmic Pole Shift Geology

by Charles Hapgood

Maps of the Ancient Sea Kings author Hapgood's classic book *Path of the Pole* is back in print! Hapgood researched Antarctica, ancient maps and geological record to conclude that the Earth's crust has slipped in the inner core many times in the past, changing the position of the pole. *Path of the P* discusses the various "pole shifts" in Earth's past, giving evidence for each one, and moves on to possible future pole shifts. Packed with illustrations, thi the sourcebook for many other books on cataclysms and pole shifts.

356 PAGES. 6X9 PAPERBACK. ILLUSTRATED. $16.95. CODE: POP.

24 hour credit card orders—call: 815-253-6390 fax: 815-253-6300
email: auphq@frontiernet.net www.adventuresunlimitedpress.com www.wexclub.com

LOST CITIES

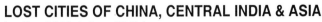

LOST CITIES OF ATLANTIS, ANCIENT EUROPE & THE MEDITERRANEAN

by David Hatcher Childress

Atlantis! The legendary lost continent comes under the close scrutiny of maverick archaeologist David Hatcher Childress in this sixth book in the internationally popular *Lost Cities* series. Childress takes the reader in search of sunken cities in the Mediterranean; across the Atlas Mountains in search of Atlantean ruins; to remote islands in search of megalithic ruins; to meet living legends and secret societies. From Ireland to Turkey, Morocco to Eastern Europe, and around the remote islands of the Mediterranean and Atlantic, Childress takes the reader on an astonishing quest for mankind's past. Ancient technology, cataclysms, megalithic construction, lost civilizations and devastating wars of the past are all explored in this book. Childress challenges the skeptics and proves that great civilizations not only existed in the past, but the modern world and its problems are reflections of the ancient world of Atlantis.

524 PAGES. 6X9 PAPERBACK. ILLUSTRATED WITH 100S OF MAPS, PHOTOS AND DIAGRAMS. BIBLIOGRAPHY & INDEX. $16.95. CODE: MED

LOST CITIES OF CHINA, CENTRAL INDIA & ASIA

by David Hatcher Childress

Like a real life "Indiana Jones," maverick archaeologist David Childress takes the reader on an incredible adventure across some of the world's oldest and most remote countries in search of lost cities and ancient mysteries. Discover ancient cities in the Gobi Desert; hear fantastic tales of lost continents, vanished civilizations and secret societies bent on ruling the world; visit forgotten monasteries in forbidding snow-capped mountains with strange tunnels to mysterious subterranean cities! A unique combination of far-out exploration and practical travel advice, it will astound and delight the experienced traveler or the armchair voyager.

429 PAGES. 6X9 PAPERBACK. ILLUSTRATED. FOOTNOTES & BIBLIOGRAPHY. $14.95. CODE: CHI

LOST CITIES OF ANCIENT LEMURIA & THE PACIFIC

by David Hatcher Childress

Was there once a continent in the Pacific? Called Lemuria or Pacifica by geologists, Mu or Pan by the mystics, there is now ample mythological, geological and archaeological evidence to "prove" that an advanced and ancient civilization once lived in the central Pacific. Maverick archaeologist and explorer David Hatcher Childress combs the Indian Ocean, Australia and the Pacific in search of the surprising truth about mankind's past. Contains photos of the underwater city on Pohnpei; explanations on how the statues were levitated around Easter Island in a clockwise vortex movement; tales of disappearing islands; Egyptians in Australia; and more.

379 PAGES. 6X9 PAPERBACK. ILLUSTRATED. FOOTNOTES & BIBLIOGRAPHY. $14.95. CODE: LEM

ANCIENT TONGA

& the Lost City of Mu'a

by David Hatcher Childress

Lost Cities series author Childress takes us to the south sea islands of Tonga, Rarotonga, Samoa and Fiji to investigate the megalithic ruins on these beautiful islands. The great empire of the Polynesians, centered on Tonga and the ancient city of Mu'a, is revealed with old photos, drawings and maps. Chapters in this book are on the Lost City of Mu'a and its many megalithic pyramids, the Ha'amonga Trilithon and ancient Polynesian astronomy, Samoa and the search for the lost land of Havai'iki, Fiji and its wars with Tonga, Rarotonga's megalithic road, and Polynesian cosmology. Material on Egyptians in the Pacific, earth changes, the fortified moat around Mu'a, lost roads, more.

218 PAGES. 6X9 PAPERBACK. ILLUSTRATED. COLOR PHOTOS. BIBLIOGRAPHY. $15.95. CODE: TONG

ANCIENT MICRONESIA

& the Lost City of Nan Madol

by David Hatcher Childress

Micronesia, a vast archipelago of islands west of Hawaii and south of Japan, contains some of the most amazing megalithic ruins in the world. Part of our *Lost Cities* series, this volume explores the incredible conformations on various Micronesian islands, especially the fantastic and little-known ruins of Nan Madol on Pohnpei Island. The huge canal city of Nan Madol contains over 250 million tons of basalt columns over an 11 square-mile area of artificial islands. Much of the huge city is submerged, and underwater structures can be found to an estimated 80 feet. Islanders' legends claim that the basalt rocks, weighing up to 50 tons, were magically levitated into place by the powerful forefathers. Other ruins in Micronesia that are profiled include the Latte Stones of the Marianas, the menhirs of Palau, the megalithic canal city on Kosrae Island, megaliths on Guam, and more.

256 PAGES. 6X9 PAPERBACK. ILLUSTRATED. INCLUDES A COLOR PHOTO SECTION. BIBLIOGRAPHY. $16.95. CODE: AMIC

24 hour credit card orders—call: 815-253-6390 fax: 815-253-6300

email: auphq@frontiernet.net www.adventuresunlimitedpress.com www.wexclub.com

LOST CITIES

TECHNOLOGY OF THE GODS

The Incredible Sciences of the Ancients

by David Hatcher Childress

Popular *Lost Cities* author David Hatcher Childress takes us into the amazing world of ancient technology, from computers in anti to the "flying machines of the gods." Childress looks at the technology that was allegedly used in Atlantis and the theory that the C Pyramid of Egypt was originally a gigantic power station. He examines tales of ancient flight and the technology that it involved; hov ancients used electricity; megalithic building techniques; the use of crystal lenses and the fire from the gods; evidence of various tech weapons in the past, including atomic weapons; ancient metallurgy and heavy machinery; the role of modern inventors suc Nikola Tesla in bringing ancient technology back into modern use; impossible artifacts; and more.

356 PAGES. 6X9 PAPERBACK. ILLUSTRATED. BIBLIOGRAPHY. $16.95. CODE: TGOD

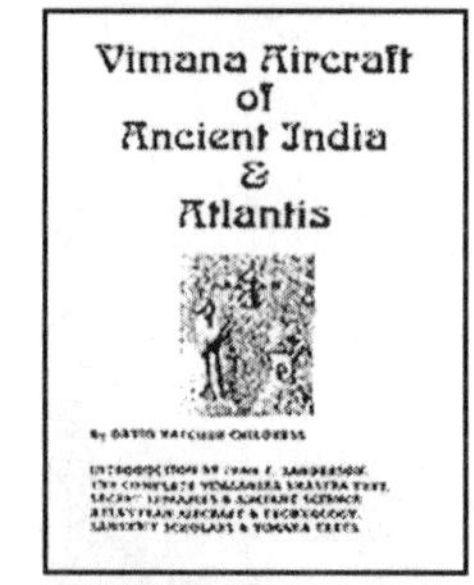

VIMANA AIRCRAFT OF ANCIENT INDIA & ATLANTIS

by David Hatcher Childress, introduction by Ivan T. Sanderson

Did the ancients have the technology of flight? In this incredible volume on ancient India, authentic Indian texts such as *Ramayana* and the *Mahabharata* are used to prove that ancient aircraft were in use more than four thousand years ago. Included in this book is the entire Fourth Century BC manuscript *Vimaanika Shastra* by the ancient author Maharishi Bharadwaaja, translated into English by the Mysore Sanskrit professor G.R. Josyer. Also included are chapters on Atlantean technology, the incredible Rama Empire of India and the devastating wars that destroyed it. Also an entire chapter on mercury vortex propulsion and mercury gyros, the power source described in the ancient Indian texts. Not to be missed by those interested in ancient civilizations or the UFO enigma.

334 PAGES. 6X9 PAPERBACK. RARE PHOTOGRAPHS, MAPS AND DRAWINGS. $15.95. CODE: VAA

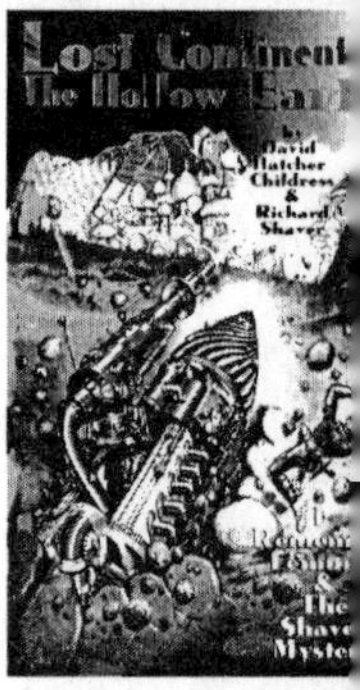

LOST CONTINENTS & THE HOLLOW EARTH

I Remember Lemuria and the Shaver Mystery

by David Hatcher Childress & Richard Shaver

Lost Continents & the Hollow Earth is Childress' thorough examination of the early hollow earth stories of Richard Shaver and the fascination that fringe fantasy subjects such as lost continents and the hollow earth have had for the American public. Shaver's rare 1948 book *I Remember Lemuria* is reprinted in its entirety, and the book is packed with illustrations from Ray Palmer's *Amazing Stories* magazine of the 1940s. Palmer and Shaver told of tunnels running through the earth—tunnels inhabited by the Deros and Teros, humanoids from an an spacefaring race that had inhabited the earth, eventually going underground, hundreds of thousands of years ago. Childress discusse famous hollow earth books and delves deep into whatever reality may be behind the stories of tunnels in the earth. Operation High J to Antarctica in 1947 and Admiral Byrd's bizarre statements, tunnel systems in South America and Tibet, the underground wor Agartha, the belief of UFOs coming from the South Pole, more.

344 PAGES. 6X9 PAPERBACK. ILLUSTRATED. $16.95. CODE: LCHE

LOST CITIES OF NORTH & CENTRAL AMERICA

by David Hatcher Childress

Down the back roads from coast to coast, maverick archaeologist and adventurer David Hatcher Childress goes deep into unkn America. With this incredible book, you will search for lost Mayan cities and books of gold, discover an ancient canal system in Ariz climb gigantic pyramids in the Midwest, explore megalithic monuments in New England, and join the astonishing quest for lost c throughout North America. From the war-torn jungles of Guatemala, Nicaragua and Honduras to the deserts, mountains and fiel Mexico, Canada, and the U.S.A., Childress takes the reader in search of sunken ruins, Viking forts, strange tunnel systems, li dinosaurs, early Chinese explorers, and fantastic lost treasure. Packed with both early and current maps, photos and illustrations.

590 PAGES. 6X9 PAPERBACK. ILLUSTRATED. FOOTNOTES & BIBLIOGRAPHY. $14.95. CODE: N

LOST CITIES & ANCIENT MYSTERIES OF SOUTH AMERICA

by David Hatcher Childress

Rogue adventurer and maverick archaeologist David Hatcher Childress takes the reader on unforgettable journeys deep into deadly jungles, high up on windswept mountains and across scorching deserts in search of lost civilizations and ancient mysteries. Travel with David and explore stone cities high in mountain forests and hear fantastic tales of Inca treasure, living dinosaurs, and a mysterious tunnel system. Whether he is hopping freight trains, searching for secret cities, or just dealing with the daily problems of food, money, and romance, the author keeps the reader spellbound. Includes both early and current maps, photos, and illustrations, and plenty of advice for the explorer planning his or her own journey of discovery.

381 PAGES. 6X9 PAPERBACK. ILLUSTRATED. FOOTNOTES & BIBLIOGRAPHY. $14.95. CODE: SAM

LOST CITIES & ANCIENT MYSTERIES OF AFRICA & ARABIA

by David Hatcher Childress

Across ancient deserts, dusty plains and steaming jungles, maverick archaeologist David Childress continues his world-wide quest for lost cities and ancient mysteries. Join him as he discovers forbidden cities in the Empty Quarter of Arabia; "Atlantean" ruins in Egypt and the Kalahari desert; a mysterious, ancient empire in the Sahara; and more. This is the tale of an extraordinary life on the road: across war-torn countries, Childress searches for King Solomon's Mines, living dinosaurs, the Ark of the Covenant and the solutions to some of the fantastic mysteries of the past.

423 PAGES. 6X9 PAPERBACK. ILLUSTRATED. FOOTNOTES & BIBLIOGRAPHY. $14.95. CODE: AFA

24 hour credit card orders—call: 815-253-6390 fax: 815-253-6300

email: auphq@frontiernet.net www.adventuresunlimitedpress.com www.wexclub.com

CONSPIRACY & HISTORY

LIQUID CONSPIRACY

JFK, LSD, the CIA, Area 51 & UFOs

by George Piccard

Underground author George Piccard on the politics of LSD, mind control, and Kennedy's involvement with Area 51 and UFOs. Reveals JFK's LSD experiences with Mary Pinchot-Meyer. The plot thickens with an ever expanding web of CIA involvement, from underground bases with UFOs seen by JFK and Marilyn Monroe (among others) to a vaster conspiracy that affects every government agency from NASA to the Justice Department. This may have been the reason that Marilyn Monroe and actress-columnist Dorothy Kilgallen were both murdered. Focusing on the bizarre side of history, *Liquid Conspiracy* takes the reader on a psychedelic tour de force. This is your government on drugs!

264 PAGES. 6X9 PAPERBACK. ILLUSTRATED. $14.95. CODE: LIQC

INSIDE THE GEMSTONE FILE

Howard Hughes, Onassis & JFK

by Kenn Thomas & David Hatcher Childress

Steamshovel Press editor Thomas takes on the Gemstone File in this run-up and run-down of the most famous underground document ever circulated. Photocopied and distributed for over 20 years, the Gemstone File is the story of Bruce Roberts, the inventor of the synthetic ruby widely used in laser technology today, and his relationship with the Howard Hughes Company and ultimately with Aristotle Onassis, the Mafia, and the CIA. Hughes kidnapped and held a drugged-up prisoner for 10 years; Onassis and his role in the Kennedy Assassination; how the Mafia ran corporate America in the 1960s; the death of Onassis' son in the crash of a small private plane in Greece; Onassis as Ian Fleming's archvillain Ernst Stavro Blofeld; more.

320 PAGES. 6X9 PAPERBACK. ILLUSTRATED. $16.00. CODE: IGF

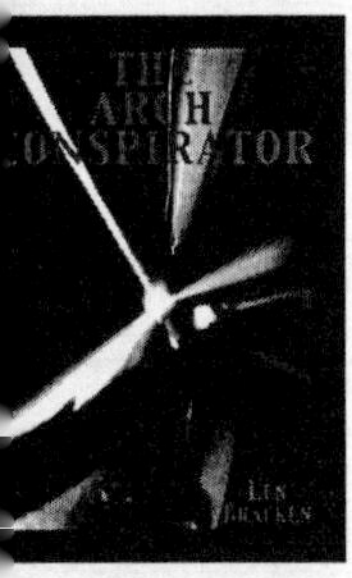

THE ARCH CONSPIRATOR

Essays and Actions

by Len Bracken

Veteran conspiracy author Len Bracken's witty essays and articles lead us down the dark corridors of conspiracy, politics, murder and mayhem. In 12 chapters Bracken takes us through a maze of interwoven tales from the Russian Conspiracy (and a few "extra notes" on conspiracies) to his interview with Costa Rican novelist Joaquin Gutierrez and his Psychogeographic Map into the Third Millennium. Other chapters in the book are A General Theory of Civil War; A False Report Exposes the Dirty Truth About South African Intelligence Services; The New-Catiline Conspiracy for the Cancellation of Debt; Anti-Labor Day; 1997 with selected Aphorisms Against Work; Solar Economics; and more. Bracken's work has appeared in such pop-conspiracy publications as *Paranoia*, *Steamshovel Press* and the *Village Voice*. Len Bracken lives in Arlington, Virginia and haunts the back alleys of Washington D.C., keeping an eye on the predators who run our country. With a gun to his head, he cranks out his rants for fringe publications and is the editor of *Extraphile*, described by *New Yorker Magazine* as "fusion conspiracy theory."

256 PAGES. 6X9 PAPERBACK. ILLUSTRATED. BIBLIOGRAPHY. $14.95. CODE: ACON.

MIND CONTROL, WORLD CONTROL

by Jim Keith

Veteran author and investigator Jim Keith uncovers a surprising amount of information on the technology, experimentation and implementation of mind control. Various chapters in this shocking book are on early CIA experiments such as Project Artichoke and Project R.H.I.C.-EDOM, the methodology and technology of implants, mind control assassins and couriers, various famous Mind Control victims such as Sirhan Sirhan and Candy Jones. Also featured in this book are chapters on how mind control technology may be linked to some UFO activity and "UFO abductions."

256 PAGES. 6X9 PAPERBACK. ILLUSTRATED. FOOTNOTES. $14.95. CODE: MCWC

NASA, NAZIS & JFK:

The Torbitt Document & the JFK Assassination

introduction by Kenn Thomas

This book emphasizes the links between "Operation Paper Clip" Nazi scientists working for NASA, the assassination of JFK, and the secret Nevada air base Area 51. The Torbitt Document also talks about the roles played in the assassination by Division Five of the FBI, the Defense Industrial Security Command (DISC), the Las Vegas mob, and the shadow corporate entities Permindex and Centro-Mondiale Commerciale. The Torbitt Document claims that the same players planned the 1962 assassination attempt on Charles de Gaul, who ultimately pulled out of NATO because he traced the "Assassination Cabal" to Permindex in Switzerland and to NATO headquarters in Brussels. The Torbitt Document paints a dark picture of NASA, the military industrial complex, and the connections to Mercury, Nevada which headquarters the "secret space program."

258 PAGES. 5X8. PAPERBACK. ILLUSTRATED. $16.00. CODE: NNJ

MIND CONTROL, OSWALD & JFK:

Were We Controlled?

introduction by Kenn Thomas

Steamshovel Press editor Kenn Thomas examines the little-known book *Were We Controlled?*, first published in 1968. The book's author, the mysterious Lincoln Lawrence, maintained that Lee Harvey Oswald was a special agent who was a mind control subject, having received an implant in 1960 at a Russian hospital. Thomas examines the evidence for implant technology and the role it could have played in the Kennedy Assassination. Thomas also looks at the mind control aspects of the RFK assassination and details the history of implant technology. A growing number of people are interested in CIA experiments and its "Silent Weapons for Quiet Wars." Looks at the case that the reporter Damon Runyon, Jr. was murdered because of this book.

256 PAGES. 6X9 PAPERBACK. ILLUSTRATED. NOTES. $16.00. CODE: MCOJ

One Adventure Place
P.O. Box 74
Kempton, Illinois 60946
United States of America
Tel.: 815-253-6390 • Fax: 815-253-6300
Email: auphq@frontiernet.net
http://www.adventuresunlimitedpress.com
or www.wexclub.com/aup

ORDERING INSTRUCTIONS

- ✓ Remit by USD$ Check, Money Order or Credit Card
- ✓ Visa, Master Card, Discover & AmEx Accepted
- ✓ Prices May Change Without Notice
- ✓ 10% Discount for 3 or more Items

SHIPPING CHARGES

United States

- ✓ Postal Book Rate { $3.00 First Item; 50¢ Each Additional Item
- ✓ Priority Mail { $4.00 First Item; $2.00 Each Additional Item
- ✓ UPS { $5.00 First Item; $1.50 Each Additional Item

NOTE: UPS Delivery Available to Mainland USA Only

Canada

- ✓ Postal Book Rate { $4.00 First Item; $1.00 Each Additional Item
- ✓ Postal Air Mail { $6.00 First Item; $2.00 Each Additional Item
- ✓ Personal Checks or Bank Drafts MUST BE USD$ and Drawn on a US Bank
- ✓ Canadian Postal Money Orders OK
- ✓ Payment MUST BE USD$

All Other Countries

- ✓ Surface Delivery { $7.00 First Item; $2.00 Each Additional Item
- ✓ Postal Air Mail { $13.00 First Item; $8.00 Each Additional Item
- ✓ Payment MUST BE USD$
- ✓ Checks and Money Orders MUST BE USD$ and Drawn on a US Bank or branch.
- ✓ Add $5.00 for Air Mail Subscription to Future *Adventures Unlimited* Catalogs

SPECIAL NOTES

- ✓ RETAILERS: Standard Discounts Available
- ✓ BACKORDERS: We Backorder all Out-of-Stock Items Unless Otherwise Requested
- ✓ PRO FORMA INVOICES: Available on Request
- ✓ VIDEOS: NTSC Mode Only. Replacement only.
- ✓ For PAL mode videos contact our other offices:

European Office:
Adventures Unlimited, Panewaal 22,
Enkhuizen, 1600 AA, The Netherlands
http: www.adventuresunlimited.nl
Check Us Out Online at:
www.adventuresunlimitedpress.com

Please check: ☑

❒ This is my first order ❒ I have ordered before ❒ This is a new addres

Name

Address

City

State/Province | Postal Code

Country

Phone day | Evening

Fax

Item Code	Item Description	Price	Qty	Total

Please check: ☑

❒ Postal-Surface

❒ Postal-Air Mail (Priority in USA)

❒ UPS (Mainland USA only)

Subtotal ➠
Less Discount-10% for 3 or more items ➠
Balance ➠
Illinois Residents 6.25% Sales Tax ➠
Previous Credit ➠
Shipping ➠
Total (check/MO in USD$ only) ➠

❒ Visa/MasterCard/Discover/Amex

Card Number

Expiration Date

10% Discount When You Order 3 or More Items!

Comments & Suggestions

Share Our Catalog with a Frienc

ANTI-GRAVITY

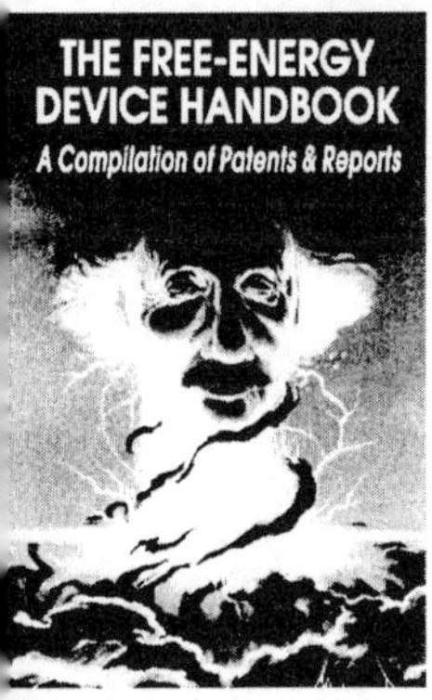

THE FREE-ENERGY DEVICE HANDBOOK

A Compilation of Patents and Reports

by David Hatcher Childress

A large-format compilation of various patents, papers, descriptions and diagrams concerning free-energy devices and systems. *The Free-Energy Device Handbook* is a visual tool for experimenters and researchers into magnetic motors and other "over-unity" devices. With chapters on the Adams Motor, the Hans Coler Generator, cold fusion, superconductors, "N" machines, space-energy generators, Nikola Tesla, T. Townsend Brown, and the latest in free-energy devices. Packed with photos, technical diagrams, patents and fascinating information, this book belongs on every science shelf. With energy and profit being a major political reason for fighting various wars, free-energy devices, if ever allowed to be mass distributed to consumers, could change the world! Get your copy now before the Department of Energy bans this book!

292 PAGES. 8X10 PAPERBACK. ILLUSTRATED. BIBLIOGRAPHY. $16.95. CODE: FEH

THE ANTI-GRAVITY HANDBOOK

edited by David Hatcher Childress, with Nikola Tesla, T.B. Paulicki, Bruce Cathie, Albert Einstein and others

The new expanded compilation of material on Anti-Gravity, Free Energy, Flying Saucer Propulsion, UFOs, Suppressed Technology, NASA Cover-ups and more. Highly illustrated with patents, technical illustrations and photos. This revised and expanded edition has more material, including photos of Area 51, Nevada, the government's secret testing facility. This classic on weird science is back in a 90s format!

- **How to build a flying saucer.**
- **Arthur C. Clarke on Anti-Gravity.**
- **Crystals and their role in levitation.**
- **Secret government research and development.**
- **Nikola Tesla on how anti-gravity airships could draw power from the atmosphere.**
- **Bruce Cathie's Anti-Gravity Equation.**
- **NASA, the Moon and Anti-Gravity.**

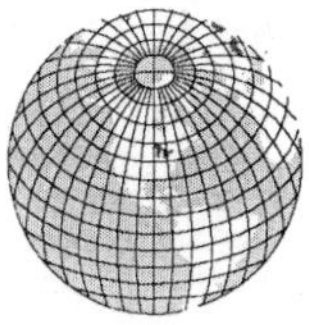

230 PAGES. 7X10 PAPERBACK. BIBLIOGRAPHY/INDEX/APPENDIX. HIGHLY ILLUSTRATED. $14.95. CODE: AGH

ANTI–GRAVITY & THE WORLD GRID

Is the earth surrounded by an intricate electromagnetic grid network offering free energy? This compilation of material on ley lines and world power points contains chapters on the geography, mathematics, and light harmonics of the earth grid. Learn the purpose of ley lines and ancient megalithic structures located on the grid. Discover how the grid made the Philadelphia Experiment possible. Explore the Coral Castle and many other mysteries, including acoustic levitation, Tesla Shields and scalar wave weaponry. Browse through the section on anti-gravity patents, and research resources.

274 PAGES. 7X10 PAPERBACK. ILLUSTRATED. $14.95. CODE: AGW

ANTI–GRAVITY & THE UNIFIED FIELD

edited by David Hatcher Childress

Is Einstein's Unified Field Theory the answer to all of our energy problems? Explored in this compilation of material is how gravity, electricity and magnetism manifest from a unified field around us. Why artificial gravity is possible; secrets of UFO propulsion; free energy; Nikola Tesla and anti-gravity airships of the 20s and 30s; flying saucers as superconducting whirls of plasma; anti-mass generators; vortex propulsion; suppressed technology; government cover-ups; gravitational pulse drive; spacecraft & more.

240 PAGES. 7X10 PAPERBACK. ILLUSTRATED. $14.95. CODE: AGU

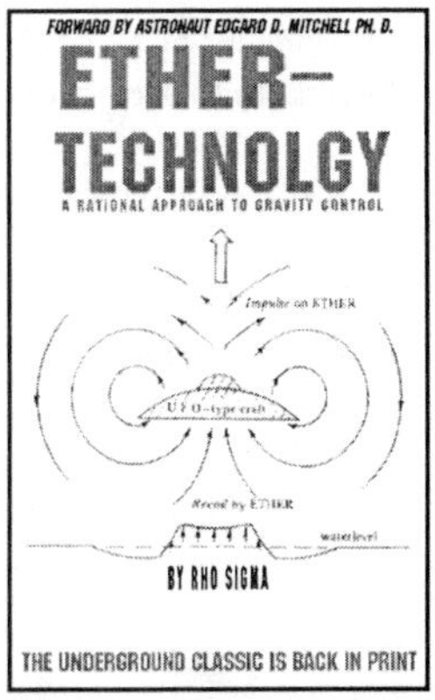

ETHER TECHNOLOGY

A Rational Approach to Gravity Control

by Rho Sigma

This classic book on anti-gravity and free energy is back in print and back in stock. Written by a well-known American scientist under the pseudonym of "Rho Sigma," this book delves into international efforts at gravity control and discoid craft propulsion. Before the Quantum Field, there was "Ether." This small, but informative book has chapters on John Searle and "Searle discs;" T. Townsend Brown and his work on anti-gravity and ether-vortex turbines. Includes a forward by former NASA astronaut Edgar Mitchell.

108 PAGES. 6X9 PAPERBACK. ILLUSTRATED. $12.95. CODE: ETT

ANTI-GRAVITY

COSMIC MATRIX

Piece for a Jig-Saw, Part Two

by Leonard G. Cramp

Leonard G. Cramp, a British aerospace engineer, wrote his first book *Space Gravity and the Flying Saucer* in 1954. Cosmi Matrix is the long-awaited sequel to his 1966 book *UFOs & Anti-Gravity: Piece for a Jig-Saw*. Cramp has had a long histor of examining UFO phenomena and has concluded that UFOs use the highest possible aeronautic science to move in the wa they do. Cramp examines anti-gravity effects and theorizes that this super-science used by the craft—described in detail in th book—can lift mankind into a new level of technology, transportation and understanding of the universe. The book takes close look at gravity control, time travel, and the interlocking web of energy between all planets in our solar system wit Leonard's unique technical diagrams. A fantastic voyage into the present and future!

364 PAGES. 6X9 PAPERBACK. ILLUSTRATED. BIBLIOGRAPHY. $16.00. CODE: CMX

UFOS AND ANTI-GRAVITY

Piece For A Jig-Saw

by Leonard G. Cramp

Leonard G. Cramp's 1966 classic book on flying saucer propulsion and suppressed technology is a highly technical look at the UFO phenomena by a trained scientist. Cramp first introduces the idea of 'anti-gravity' and introduces us to the various theories of gravitation. He then examines the technology necessary to build a flying saucer and examines in great detail the technical aspects of such a craft. Cramp's book is a wealth of material and diagrams on flying saucers, anti-gravity, suppressed technology, G-fields and UFOs. Chapters include Crossroads of Aerodymanics, Aerodynamic Saucers, Limitations of Rocketry, Gravitation and the Ether, Gravitational Spaceships, G-Field Lift Effects, The Bi-Field Theory, VTOL and Hovercraft, Analysis of UFO photos, more.

388 PAGES. 6X9 PAPERBACK. ILLUSTRATED. $16.95. CODE: UAG

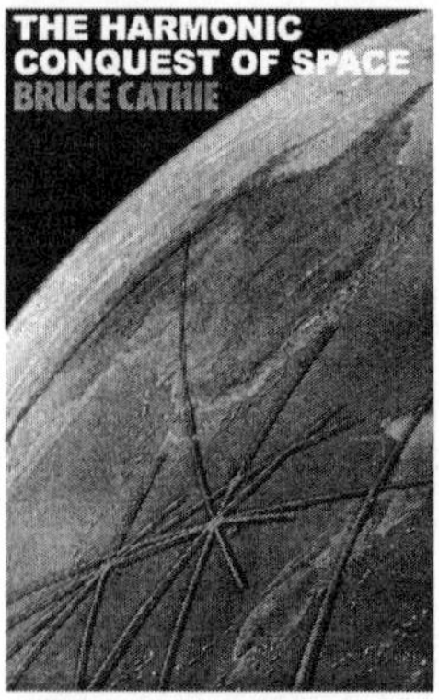

THE HARMONIC CONQUEST OF SPACE

by Captain Bruce Cathie

Chapters include: Mathematics of the World Grid; the Harmonics of Hiroshima and Nagasaki; Harmonic Transmission and Receiving; the Link Between Human Brain Waves; the Cavity Resonance between the Earth; the Ionosphere and Gravity; Edgar Cayce—the Harmonics of the Subconscious; Stonehenge; the Harmonics of the Moon; the Pyramids of Mars; Nikola Tesla's Electric Car; the Robert Adam Pulsed Electric Motor Generator; Harmonic Clues to the Unified Field; and more. Also included are tables showing th harmonic relations between the earth's magnetic field, the speed of light, and anti-gravity/gravity acceleration at differer points on the earth's surface. New chapters in this edition on the giant stone spheres of Costa Rica, Atomic Tests and Volcani Activity, and a chapter on Ayers Rock analysed with Stone Mountain, Georgia.

248 PAGES. 6X9. PAPERBACK. ILLUSTRATED. BIBLIOGRAPHY. $16.95. CODE: HCS

THE ENERGY GRID

Harmonic 695, The Pulse of the Universe

by Captain Bruce Cathie.

This is the breakthrough book that explores the incredible potential of the Energy Grid and the Earth's Unified Field all arou us. Cathie's first book, *Harmonic 33*, was published in 1968 when he was a commercial pilot in New Zealand. Since the Captain Bruce Cathie has been the premier investigator into the amazing potential of the infinite energy that surrounds o planet every microsecond. Cathie investigates the Harmonics of Light and how the Energy Grid is created. In this amazi book are chapters on UFO Propulsion, Nikola Tesla, Unified Equations, the Mysterious Aerials, Pythagoras & the Grid, Nucle Detonation and the Grid, Maps of the Ancients, an Australian Stonehenge examined, more.

255 PAGES. 6X9 TRADEPAPER. ILLUSTRATED. $15.95. CODE: TEG

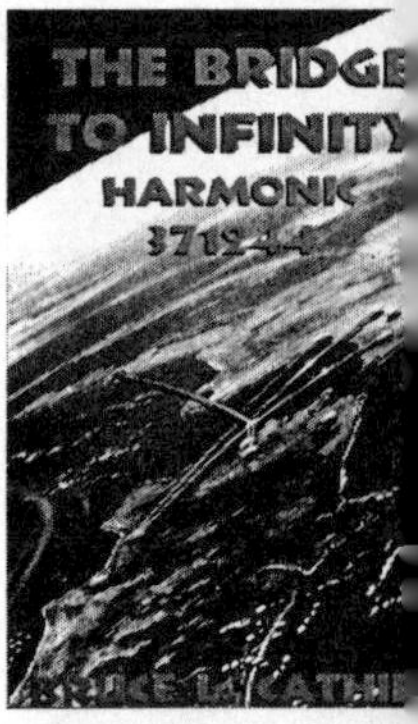

THE BRIDGE TO INFINITY

Harmonic 371244

by Captain Bruce Cathie

Cathie has popularized the concept that the earth is crisscrossed by an electromagnetic grid system that can be used for anti-gravity, free energy, levitation and more. The book includes a new analysis of the harmonic nature of reality, acoustic levitation, pyramid power, harmonic receiver towers and UFO propulsion. It concludes that today's scientists have at their command a fantastic store of knowledge with which to advance the welfare of the human race.

204 PAGES. 6X9 TRADEPAPER. ILLUSTRATED. $14.95. CODE: BTF

MAN-MADE UFOS 1944—1994

Fifty Years of Suppression

by Renato Vesco & David Hatcher Childress

A comprehensive look at the early "flying saucer" technology of Nazi Germany and the genesis of man-made UFOs. This book takes us from the work of captured German scientists to escaped battalions of Germans, secr communities in South America and Antarctica to todays state-of-the-art "Dreamland" flying machines. Heavily illustrate this astonishing book blows the lid off the "government UFO conspiracy" and explains with technical diagrams the techno ogy involved. Examined in detail are secret underground airfields and factories; German secret weapons; "suction" aircra the origin of NASA; gyroscopic stabilizers and engines; the secret Marconi aircraft factory in South America; and mo Introduction by W.A. Harbinson, author of the Dell novels *GENESIS* and *REVELATION*.

318 PAGES. 6X9 PAPERBACK. ILLUSTRATED. INDEX & FOOTNOTES. $18.95. CODE: MMU

TESLA TECHNOLOGY

THE FANTASTIC INVENTIONS OF NIKOLA TESLA

Nikola Tesla with additional material by David Hatcher Childress

This book is a readable compendium of patents, diagrams, photos and explanations of the many incredible inventions of the originator of the modern era of electrification. In Tesla's own words are such topics as wireless transmission of power, death rays, and radio-controlled airships. In addition, rare material on German bases in Antarctica and South America, and a secret city built at a remote jungle site in South America by one of Tesla's students, Guglielmo Marconi. Marconi's secret group claims to have built flying saucers in the 1940s and to have gone to Mars in the early 1950s! Incredible photos of these Tesla craft are included. The Ancient Atlantean system of broadcasting energy through a grid system of obelisks and pyramids is discussed, and a fascinating concept comes out of one chapter: that Egyptian engineers had to wear protective metal head-shields while in these power plants, hence the Egyptian Pharoah's head covering as well as the Face on Mars!

•His plan to transmit free electricity into the atmosphere. •How electrical devices would work using only small antennas mounted on them. •Why unlimited power could be utilized anywhere on earth. •How radio and radar technology can be used as death-ray weapons in Star Wars. •Includes an appendix of Supreme Court documents on dismantling his free energy towers. •Tesla's Death Rays, Ozone generators, and more…

342 PAGES. 6X9 PAPERBACK. ILLUSTRATED. BIBLIOGRAPHY AND APPENDIX. $16.95. CODE: FINT

THE TESLA PAPERS

Nikola Tesla on Free Energy & Wireless Transmission of Power

by Nikola Tesla, edited by David Hatcher Childress

In the tradition of *The Fantastic Inventions of Nikola Tesla*, *The Anti-Gravity Handbook* and *The Free-Energy Device Handbook*, science and UFO author David Hatcher Childress takes us into the incredible world of Nikola Tesla and his amazing inventions. Tesla's rare article "The Problem of Increasing Human Energy with Special Reference to the Harnessing of the Sun's Energy" is included. This lengthy article was originally published in the June 1900 issue of *The Century Illustrated Monthly Magazine* and it was the outline for Tesla's master blueprint for the world. Tesla's fantastic vision of the future, including wireless power, anti-gravity, free energy and highly advanced solar power.

Also included are some of the papers, patents and material collected on Tesla at the Colorado Springs Tesla Symposiums, including papers on:

•The Secret History of Wireless Transmission •Tesla and the Magnifying Transmitter
•Design and Construction of a half-wave Tesla Coil •Electrostatics: A Key to Free Energy
•Progress in Zero-Point Energy Research •Electromagnetic Energy from Antennas to Atoms
•Tesla's Particle Beam Technology •Fundamental Excitatory Modes of the Earth-Ionosphere Cavity

325 PAGES. 8X10 PAPERBACK. ILLUSTRATED. $16.95. CODE: TTP

LOST SCIENCE

by Gerry Vassilatos

Secrets of Cold War Technology author Vassilatos on the remarkable lives, astounding discoveries, and incredible inventions of such famous people as Nikola Tesla, Dr. Royal Rife, T.T. Brown, and T. Henry Moray. Read about the aura research of Baron Karl von Reichenbach, the wireless technology of Antonio Meucci, the controlled fusion devices of Philo Farnsworth, the earth battery of Nathan Stubblefield, and more. What were the twisted intrigues which surrounded the often deliberate attempts to stop this technology? Vassilatos claims that we are living hundreds of years behind our intended level of technology and we must recapture this "lost science."

304 PAGES. 6X9 PAPERBACK. ILLUSTRATED. BIBLIOGRAPHY. $16.95. CODE: LOS

SECRETS OF COLD WAR TECHNOLOGY

Project HAARP and Beyond

by Gerry Vassilatos

Vassilatos reveals that "Death Ray" technology has been secretly researched and developed since the turn of the century. Included are chapters on such inventors and their devices as H.C. Vion, the developer of auroral energy receivers; Dr. Selim Lemstrom's pre-Tesla experiments; the early beam weapons of Grindell-Mathews, Ulivi, Turpain and others; John Hettenger and his early beam power systems. Learn about Project Argus, Project Teak and Project Orange; EMP experiments in the 60s; why the Air Force directed the construction of a huge Ionospheric "backscatter" telemetry system across the Pacific just after WWII; why Raytheon has collected every patent relevant to HAARP over the past few years; more.

250 PAGES. 6X9 PAPERBACK. ILLUSTRATED. $15.95. CODE: SCWT

HAARP

The Ultimate Weapon of the Conspiracy

by Jerry Smith

The HAARP project in Alaska is one of the most controversial projects ever undertaken by the U.S. Government. Jerry Smith gives us the history of the HAARP project and explains how works, in technically correct yet easy to understand language. At best, HAARP is science out-of-control; at worst, HAARP could be the most dangerous device ever created, a futuristic technology that is everything from super-beam weapon to world-wide mind control device. Topics include Over-the-Horizon Radar and HAARP, Mind Control, ELF and HAARP, The Telsa Connection, The Russian Woodpecker, GWEN & HAARP, Earth Penetrating Tomography, Weather Modification, Secret Science of the Conspiracy, more. Includes the complete 1987 Eastlund patent for his pulsed super-weapon that he claims was stolen by the HAARP Project.

256 PAGES. 6X9 PAPERBACK. ILLUSTRATED. $14.95. CODE: HARP

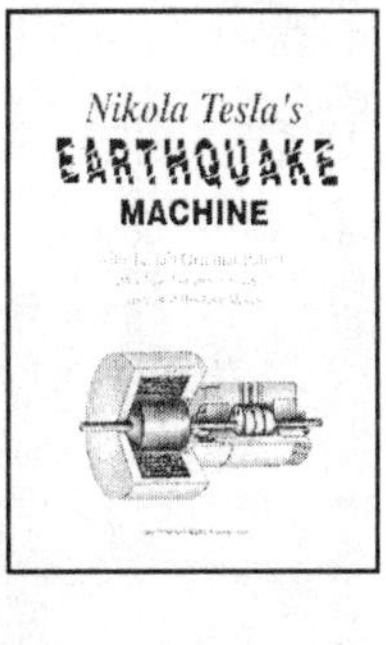

NIKOLA TESLA'S EARTHQUAKE MACHINE

with Tesla's Original Patents

by Dale Pond and Walter Baumgartner

Now, for the first time, the secrets of Nikola Tesla's Earthquake Machine are available. Although this book discusses in detail Nikola Tesla's 1894 "Earthquake Oscillator," it is also about the new technology of sonic vibrations which produce a resonance effect that can be used to cause earthquakes. Discussed are Tesla Oscillators, Vibration Physics, Amplitude Modulated Additive Synthesis, Tele-Geo-dynamics, Solar Heat Pump Apparatus, Vortex Tube Coolers, the Serogodsky Motor, more. Plenty of technical diagrams. Be the first on your block to have a Tesla Earthquake Machine!

175 PAGES. 9X11 PAPERBACK. ILLUSTRATED. BIBLIOGRAPHY & INDEX. $16.95. CODE: TEM

STRANGE SCIENCE

UNDERGROUND BASES & TUNNELS

What is the Government Trying to Hide?

by Richard Sauder, Ph.D.

Working from government documents and corporate records, Sauder has compiled an impressive book that digs below the surface of the military's super-secret underground! Go behind the scenes into little-known corners of the public record and discover how corporate America has worked hand-in-glove with the Pentagon for decades, dreaming about, planning, and actually constructing, secret underground bases. This book includes chapters on the locations of the bases, the tunneling technology, various military designs for underground bases, nuclear testing & underground bases, abductions, needles & implants, military involvement in "alien" cattle mutilations, more. 50 page photo & map insert.

201 PAGES. 6X9 PAPERBACK. ILLUSTRATED. $15.95. CODE: UGB

UNDERWATER & UNDERGROUND BASES

Surprising Facts the Government Does Not Want You to Know

by Richard Sauder

Dr. Richard Sauder's brand new book *Underwater and Underground Bases* is an explosive, eye-opening sequel to his best-selling, *Undergrou Bases and Tunnels: What is the Government Trying to Hide?* Dr. Sauder lays out the amazing evidence and government paper trail for the constru tion of huge, manned bases offsore, in mid-ocean, and deep beneath the sea floor! Bases big enough to secretly dock submarines! Official Uni States Navy documents, and other hard evidence, raise many questions about what really lies 20,000 leagues beneath the sea. Many UFOs have be seen coming and going from the world's oceans, seas and lakes, implying the existence of secret underwater bases. Hold on to your hats: Jules Ve may not have been so far from the truth, after all! Dr. Sauder also adds to his incredible database of underground bases onshore. New, breakthrou material reveals the existence of additional clandestine underground facilities as well as the surprising location of one of the CIA's own undergrou bases. Plus, new information on tunneling and cutting-edge, high speed rail magnetic-levitation (MagLev) technology. There are many rumors secret, underground tunnels with MagLev trains hurtling through them. Is there truth behind the rumors? *Underwater and Underground Ba* carefully examines the evidence and comes to a thought provoking conclusion!

264 PAGES. 6X9 PAPERBACK. ILLUSTRATED. BIBLIOGRAPHY. INDEX. $16.95. CODE: UUB

KUNDALINI TALES

by Richard Sauder, Ph.D.

Underground Bases and Tunnels author Richard Sauder's second book on his personal experiences and provocative research into sp taneous spiritual awakening, out-of-body journeys, encounters with secretive governmental powers, daylight sightings of UFOs, more. Sauder continues his studies of underground bases with new information on the occult underpinnings of the U.S. space progr The book also contains a breakthrough section that examines actual U.S. patents for devices that manipulate minds and thoughts fro remote distance. Included are chapters on the secret space program and a 130-page appendix of patents and schematic diagrams of se technology and mind control devices.

296 PAGES. 7X10 PAPERBACK. ILLUSTRATED. BIBLIOGRAPHY. $14.95. CODE: KTAL

THE TIME TRAVEL HANDBOOK

A Manual of Practical Teleportation & Time Travel

edited by David Hatcher Childress

In the tradition of *The Anti-Gravity Handbook* and *The Free-Energy Device Handbook,* science and UFO author David Hatcher Childre takes us into the weird world of time travel and teleportation. Not just a whacked-out look at science fiction, this book is an authoritati chronicling of real-life time travel experiments, teleportation devices and more. *The Time Travel Handbook* takes the reader beyond t government experiments and deep into the uncharted territory of early time travellers such as Nikola Tesla and Guglielmo Marconi and th alleged time travel experiments, as well as the Wilson Brothers of EMI and their connection to the Philadelphia Experiment—the U.S. Nav forays into invisibility, time travel, and teleportation. Childress looks into the claims of time travelling individuals, and investigates the unusu claim that the pyramids on Mars were built in the future and sent back in time. A highly visual, large format book, with patents, photos a schematics. Be the first on your block to build your own time travel device!

316 PAGES. 7X10 PAPERBACK. ILLUSTRATED. $16.95. CODE: TTH

MAPS OF THE ANCIENT SEA KINGS

Evidence of Advanced Civilization in the Ice Age

by Charles H. Hapgood

Charles Hapgood's classic 1966 book on ancient maps produces concrete evidence of an advanced world-wide civilization existing many thousands of years before ancient Egypt. He has found the evidence in the Piri Reis Map that shows Antarctica, the Hadji Ahmed map, the Oronteus Finaeus and other amazing maps. Hapgood concluded that these maps were made from more ancient maps from the various ancient archives around the world, now lost. Not only were these unknown people more advanced in mapmaking than any people prior to the 18th century, it appears they mapped all the continents. The Americas were mapped thousands of years before Columbus. Antarctica was mapped when its coasts were free of ice.

316 PAGES. 7X10 PAPERBACK. ILLUSTRATED. BIBLIOGRAPHY & INDEX. $19.95. CODE: MASK

PATH OF THE POLE

Cataclysmic Pole Shift Geology

by Charles Hapgood

Maps of the Ancient Sea Kings author Hapgood's classic book *Path of the Pole* is back in print! Hapgood researched Antarctica, ancient maps and geological record to conclude that the Earth's crust has slipped in the inner core many times in the past, changing the position of the pole. *Path of the P* discusses the various "pole shifts" in Earth's past, giving evidence for each one, and moves on to possible future pole shifts. Packed with illustrations, thi the sourcebook for many other books on cataclysms and pole shifts.

356 PAGES. 6X9 PAPERBACK. ILLUSTRATED. $16.95. CODE: POP.

LOST CITIES

LOST CITIES OF ATLANTIS, ANCIENT EUROPE & THE MEDITERRANEAN

by David Hatcher Childress

Atlantis! The legendary lost continent comes under the close scrutiny of maverick archaeologist David Hatcher Childress in this sixth book in the internationally popular *Lost Cities* series. Childress takes the reader in search of sunken cities in the Mediterranean; across the Atlas Mountains in search of Atlantean ruins; to remote islands in search of megalithic ruins; to meet living legends and secret societies. From Ireland to Turkey, Morocco to Eastern Europe, and around the remote islands of the Mediterranean and Atlantic, Childress takes the reader on an astonishing quest for mankind's past. Ancient technology, cataclysms, megalithic construction, lost civilizations and devastating wars of the past are all explored in this book. Childress challenges the skeptics and proves that great civilizations not only existed in the past, but the modern world and its problems are reflections of the ancient world of Atlantis.

524 PAGES. 6X9 PAPERBACK. ILLUSTRATED WITH 100S OF MAPS, PHOTOS AND DIAGRAMS. BIBLIOGRAPHY & INDEX. $16.95. CODE: MED

LOST CITIES OF CHINA, CENTRAL INDIA & ASIA

by David Hatcher Childress

Like a real life "Indiana Jones," maverick archaeologist David Childress takes the reader on an incredible adventure across some of the world's oldest and most remote countries in search of lost cities and ancient mysteries. Discover ancient cities in the Gobi Desert; hear fantastic tales of lost continents, vanished civilizations and secret societies bent on ruling the world; visit forgotten monasteries in forbidding snow-capped mountains with strange tunnels to mysterious subterranean cities! A unique combination of far-out exploration and practical travel advice, it will astound and delight the experienced traveler or the armchair voyager.

429 PAGES. 6X9 PAPERBACK. ILLUSTRATED. FOOTNOTES & BIBLIOGRAPHY. $14.95. CODE: CHI

LOST CITIES OF ANCIENT LEMURIA & THE PACIFIC

by David Hatcher Childress

Was there once a continent in the Pacific? Called Lemuria or Pacifica by geologists, Mu or Pan by the mystics, there is now ample mythological, geological and archaeological evidence to "prove" that an advanced and ancient civilization once lived in the central Pacific. Maverick archaeologist and explorer David Hatcher Childress combs the Indian Ocean, Australia and the Pacific in search of the surprising truth about mankind's past. Contains photos of the underwater city on Pohnpei; explanations on how the statues were levitated around Easter Island in a clockwise vortex movement; tales of disappearing islands; Egyptians in Australia; and more.

379 PAGES. 6X9 PAPERBACK. ILLUSTRATED. FOOTNOTES & BIBLIOGRAPHY. $14.95. CODE: LEM

ANCIENT TONGA

& the Lost City of Mu'a

by David Hatcher Childress

Lost Cities series author Childress takes us to the south sea islands of Tonga, Rarotonga, Samoa and Fiji to investigate the megalithic ruins on these beautiful islands. The great empire of the Polynesians, centered on Tonga and the ancient city of Mu'a, is revealed with old photos, drawings and maps. Chapters in this book are on the Lost City of Mu'a and its many megalithic pyramids, the Ha'amonga Trilithon and ancient Polynesian astronomy, Samoa and the search for the lost land of Havai'iki, Fiji and its wars with Tonga, Rarotonga's megalithic road, and Polynesian cosmology. Material on Egyptians in the Pacific, earth changes, the fortified moat around Mu'a, lost roads, more.

218 PAGES. 6X9 PAPERBACK. ILLUSTRATED. COLOR PHOTOS. BIBLIOGRAPHY. $15.95. CODE: TONG

ANCIENT MICRONESIA

& the Lost City of Nan Madol

by David Hatcher Childress

Micronesia, a vast archipelago of islands west of Hawaii and south of Japan, contains some of the most amazing megalithic ruins in the world. Part of our *Lost Cities* series, this volume explores the incredible conformations on various Micronesian islands, especially the fantastic and little-known ruins of Nan Madol on Pohnpei Island. The huge canal city of Nan Madol contains over 250 million tons of basalt columns over an 11 square-mile area of artificial islands. Much of the huge city is submerged, and underwater structures can be found to an estimated 80 feet. Islanders' legends claim that the basalt rocks, weighing up to 50 tons, were magically levitated into place by the powerful forefathers. Other ruins in Micronesia that are profiled include the Latte Stones of the Marianas, the menhirs of Palau, the megalithic canal city on Kosrae Island, megaliths on Guam, and more.

256 PAGES. 6X9 PAPERBACK. ILLUSTRATED. INCLUDES A COLOR PHOTO SECTION. BIBLIOGRAPHY. $16.95. CODE: AMIC

LOST CITIES

TECHNOLOGY OF THE GODS
The Incredible Sciences of the Ancients
by David Hatcher Childress

Popular *Lost Cities* author David Hatcher Childress takes us into the amazing world of ancient technology, from computers in antic to the "flying machines of the gods." Childress looks at the technology that was allegedly used in Atlantis and the theory that the G Pyramid of Egypt was originally a gigantic power station. He examines tales of ancient flight and the technology that it involved; how ancients used electricity; megalithic building techniques; the use of crystal lenses and the fire from the gods; evidence of various h tech weapons in the past, including atomic weapons; ancient metallurgy and heavy machinery; the role of modern inventors suc Nikola Tesla in bringing ancient technology back into modern use; impossible artifacts; and more.

356 PAGES. 6X9 PAPERBACK. ILLUSTRATED. BIBLIOGRAPHY. $16.95. CODE: TGOD

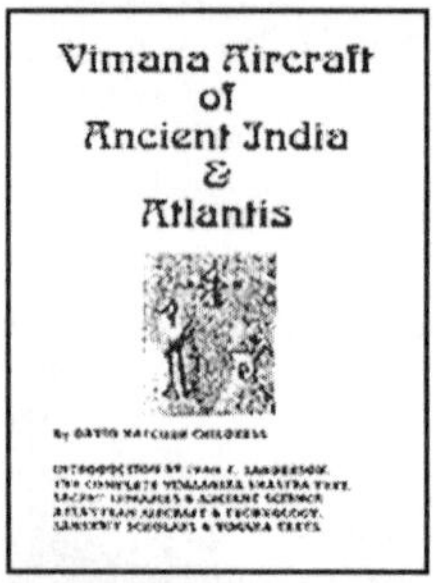

VIMANA AIRCRAFT OF ANCIENT INDIA & ATLANTIS
by David Hatcher Childress, introduction by Ivan T. Sanderson

Did the ancients have the technology of flight? In this incredible volume on ancient India, authentic Indian texts such as *Ramayana* and the *Mahabharata* are used to prove that ancient aircraft were in use more than four thousand years ago. Included in this book is the entire Fourth Century BC manuscript *Vimaanika Shastra* by the ancient author Maharishi Bharadwaaja, translated into English by the Mysore Sanskrit professor G.R. Josyer. Also included are chapters on Atlantean technology, the incredible Rama Empire of India and the devastating wars that destroyed it. Also an entire chapter on mercury vortex propulsion and mercury gyros, the power source described in the ancient Indian texts. Not to be missed by those interested in ancient civilizations or the UFO enigma.

334 PAGES. 6X9 PAPERBACK. RARE PHOTOGRAPHS, MAPS AND DRAWINGS. $15.95. CODE: VAA

LOST CONTINENTS & THE HOLLOW EARTH
I Remember Lemuria and the Shaver Mystery
by David Hatcher Childress & Richard Shaver

Lost Continents & the Hollow Earth is Childress' thorough examination of the early hollow earth stories of Richard Shaver and the fascination that fringe fantasy subjects such as lost continents and the hollow earth have had for the American public. Shaver's rare 1948 book *I Remember Lemuria* is reprinted in its entirety, and the book is packed with illustrations from Ray Palmer's *Amazing Stories* magazine of the 1940s. Palmer and Shaver told of tunnels running through the earth—tunnels inhabited by the Deros and Teros, humanoids from an anc spacefaring race that had inhabited the earth, eventually going underground, hundreds of thousands of years ago. Childress discusses famous hollow earth books and delves deep into whatever reality may be behind the stories of tunnels in the earth. Operation High J to Antarctica in 1947 and Admiral Byrd's bizarre statements, tunnel systems in South America and Tibet, the underground worl Agartha, the belief of UFOs coming from the South Pole, more.

344 PAGES. 6X9 PAPERBACK. ILLUSTRATED. $16.95. CODE: LCHE

LOST CITIES OF NORTH & CENTRAL AMERICA
by David Hatcher Childress

Down the back roads from coast to coast, maverick archaeologist and adventurer David Hatcher Childress goes deep into unkn America. With this incredible book, you will search for lost Mayan cities and books of gold, discover an ancient canal system in Ariz climb gigantic pyramids in the Midwest, explore megalithic monuments in New England, and join the astonishing quest for lost c throughout North America. From the war-torn jungles of Guatemala, Nicaragua and Honduras to the deserts, mountains and fielc Mexico, Canada, and the U.S.A., Childress takes the reader in search of sunken ruins, Viking forts, strange tunnel systems, li dinosaurs, early Chinese explorers, and fantastic lost treasure. Packed with both early and current maps, photos and illustrations.

590 PAGES. 6X9 PAPERBACK. ILLUSTRATED. FOOTNOTES & BIBLIOGRAPHY. $14.95. CODE: N

LOST CITIES & ANCIENT MYSTERIES OF SOUTH AMERICA
by David Hatcher Childress

Rogue adventurer and maverick archaeologist David Hatcher Childress takes the reader on unforgettable journeys deep into deadly jungles, high up on windswept mountains and across scorching deserts in search of lost civilizations and ancient mysteries. Travel with David and explore stone cities high in mountain forests and hear fantastic tales of Inca treasure, living dinosaurs, and a mysterious tunnel system. Whether he is hopping freight trains, searching for secret cities, or just dealing with the daily problems of food, money, and romance, the author keeps the reader spellbound. Includes both early and current maps, photos, and illustrations, and plenty of advice for the explorer planning his or her own journey of discovery.

381 PAGES. 6X9 PAPERBACK. ILLUSTRATED. FOOTNOTES & BIBLIOGRAPHY. $14.95. CODE: SAM

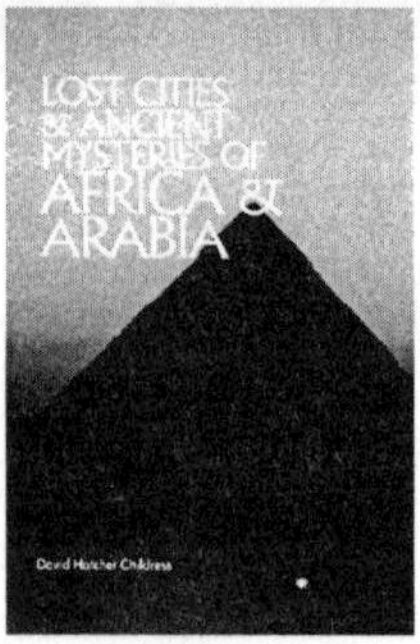

LOST CITIES & ANCIENT MYSTERIES OF AFRICA & ARABIA
by David Hatcher Childress

Across ancient deserts, dusty plains and steaming jungles, maverick archaeologist David Childress continues his world-wide quest for lost cities and ancient mysteries. Join him as he discovers forbidden cities in the Empty Quarter of Arabia; "Atlantean" ruins in Egypt and the Kalahari desert; a mysterious, ancient empire in the Sahara; and more. This is the tale of an extraordinary life on the road: across war-torn countries, Childress searches for King Solomon's Mines, living dinosaurs, the Ark of the Covenant and the solutions to some of the fantastic mysteries of the past.

423 PAGES. 6X9 PAPERBACK. ILLUSTRATED. FOOTNOTES & BIBLIOGRAPHY. $14.95. CODE: AFA

24 hour credit card orders—call: 815-253-6390 fax: 815-253-6300

email: auphq@frontiernet.net www.adventuresunlimitedpress.com www.wexclub.com

CONSPIRACY & HISTORY

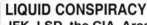

LIQUID CONSPIRACY

JFK, LSD, the CIA, Area 51 & UFOs

by George Piccard

Underground author George Piccard on the politics of LSD, mind control, and Kennedy's involvement with Area 51 and UFOs. Reveals JFK's LSD experiences with Mary Pinchot-Meyer. The plot thickens with an ever expanding web of CIA involvement, from underground bases with UFOs seen by JFK and Marilyn Monroe (among others) to a vaster conspiracy that affects every government agency from NASA to the Justice Department. This may have been the reason that Marilyn Monroe and actress-columnist Dorothy Kilgallen were both murdered. Focusing on the bizarre side of history, *Liquid Conspiracy* takes the reader on a psychedelic tour de force. This is your government on drugs!

264 PAGES. 6X9 PAPERBACK. ILLUSTRATED. $14.95. CODE: LIQC

INSIDE THE GEMSTONE FILE

Howard Hughes, Onassis & JFK

by Kenn Thomas & David Hatcher Childress

Steamshovel Press editor Thomas takes on the Gemstone File in this run-up and run-down of the most famous underground document ever circulated. Photocopied and distributed for over 20 years, the Gemstone File is the story of Bruce Roberts, the inventor of the synthetic ruby widely used in laser technology today, and his relationship with the Howard Hughes Company and ultimately with Aristotle Onassis, the Mafia, and the CIA. Hughes kidnapped and held a drugged-up prisoner for 10 years; Onassis and his role in the Kennedy Assassination; how the Mafia ran corporate America in the 1960s; the death of Onassis' son in the crash of a small private plane in Greece; Onassis as Ian Fleming's archvillain Ernst Stavro Blofeld; more.

320 PAGES. 6X9 PAPERBACK. ILLUSTRATED. $16.00. CODE: IGF

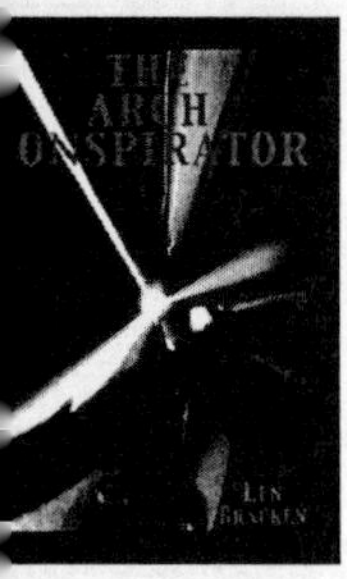

THE ARCH CONSPIRATOR

Essays and Actions

by Len Bracken

Veteran conspiracy author Len Bracken's witty essays and articles lead us down the dark corridors of conspiracy, politics, murder and mayhem. In 12 chapters Bracken takes us through a maze of interwoven tales from the Russian Conspiracy (and a few "extra notes" on conspiracies) to his interview with Costa Rican novelist Joaquin Gutierrez and his Psychogeographic Map into the Third Millennium. Other chapters in the book are A General Theory of Civil War; A False Report Exposes the Dirty Truth About South African Intelligence Services; The New-Catiline Conspiracy for the Cancellation of Debt; Anti-Labor Day; 1997 with selected Aphorisms Against Work; Solar Economics; and more. Bracken's work has appeared in such pop-conspiracy publications as *Paranoia*, *Steamshovel Press* and the *Village Voice*. Len Bracken lives in Arlington, Virginia and haunts the back alleys of Washington D.C., keeping an eye on the predators who run our country. With a gun to his head, he cranks out his rants for fringe publications and is the editor of *Extraphile*, described by *New Yorker Magazine* as "fusion conspiracy theory."

256 PAGES. 6X9 PAPERBACK. ILLUSTRATED. BIBLIOGRAPHY. $14.95. CODE: ACON.

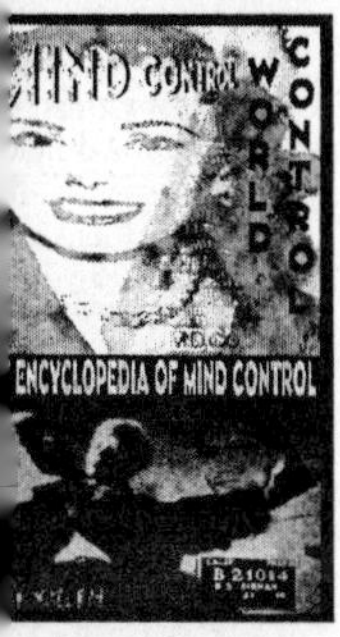

MIND CONTROL, WORLD CONTROL

by Jim Keith

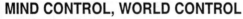

Veteran author and investigator Jim Keith uncovers a surprising amount of information on the technology, experimentation and implementation of mind control. Various chapters in this shocking book are on early CIA experiments such as Project Artichoke and Project R.H.I.C.-EDOM, the methodology and technology of implants, mind control assassins and couriers, various famous Mind Control victims such as Sirhan Sirhan and Candy Jones. Also featured in this book are chapters on how mind control technology may be linked to some UFO activity and "UFO abductions."

256 PAGES. 6X9 PAPERBACK. ILLUSTRATED. FOOTNOTES. $14.95. CODE: MCWC

NASA, NAZIS & JFK:

The Torbitt Document & the JFK Assassination

introduction by Kenn Thomas

This book emphasizes the links between "Operation Paper Clip" Nazi scientists working for NASA, the assassination of JFK, and the secret Nevada air base Area 51. The Torbitt Document also talks about the roles played in the assassination by Division Five of the FBI, the Defense Industrial Security Command (DISC), the Las Vegas mob, and the shadow corporate entities Permindex and Centro-Mondiale Commerciale. The Torbitt Document claims that the same players planned the 1962 assassination attempt on Charles de Gaul, who ultimately pulled out of NATO because he traced the "Assassination Cabal" to Permindex in Switzerland and to NATO headquarters in Brussels. The Torbitt Document paints a dark picture of NASA, the military industrial complex, and the connections to Mercury, Nevada which headquarters the "secret space program."

258 PAGES. 5X8. PAPERBACK. ILLUSTRATED. $16.00. CODE: NNJ

MIND CONTROL, OSWALD & JFK:

Were We Controlled?

introduction by Kenn Thomas

Steamshovel Press editor Kenn Thomas examines the little-known book *Were We Controlled?*, first published in 1968. The book's author, the mysterious Lincoln Lawrence, maintained that Lee Harvey Oswald was a special agent who was a mind control subject, having received an implant in 1960 at a Russian hospital. Thomas examines the evidence for implant technology and the role it could have played in the Kennedy Assassination. Thomas also looks at the mind control aspects of the RFK assassination and details the history of implant technology. A growing number of people are interested in CIA experiments and its "Silent Weapons for Quiet Wars." Looks at the case that the reporter Damon Runyon, Jr. was murdered because of this book.

256 PAGES. 6X9 PAPERBACK. ILLUSTRATED. NOTES. $16.00. CODE: MCOJ

One Adventure Place
P.O. Box 74
Kempton, Illinois 60946
United States of America
Tel.: 815-253-6390 • Fax: 815-253-6300
Email: auphq@frontiernet.net
http://www.adventuresunlimitedpress.com
or www.wexclub.com/aup

ORDERING INSTRUCTIONS

- ✓ Remit by USD$ Check, Money Order or Credit Card
- ✓ Visa, Master Card, Discover & AmEx Accepted
- ✓ Prices May Change Without Notice
- ✓ 10% Discount for 3 or more Items

SHIPPING CHARGES

United States

- ✓ Postal Book Rate { $3.00 First Item; 50¢ Each Additional Item
- ✓ Priority Mail { $4.00 First Item; $2.00 Each Additional Item
- ✓ UPS { $5.00 First Item; $1.50 Each Additional Item

NOTE: UPS Delivery Available to Mainland USA Only

Canada

- ✓ Postal Book Rate { $4.00 First Item; $1.00 Each Additional Item
- ✓ Postal Air Mail { $6.00 First Item; $2.00 Each Additional Item
- ✓ Personal Checks or Bank Drafts MUST BE USD$ and Drawn on a US Bank
- ✓ Canadian Postal Money Orders OK
- ✓ Payment MUST BE USD$

All Other Countries

- ✓ Surface Delivery { $7.00 First Item; $2.00 Each Additional Item
- ✓ Postal Air Mail { $13.00 First Item; $8.00 Each Additional Item
- ✓ Payment MUST BE USD$
- ✓ Checks and Money Orders MUST BE USD$ and Drawn on a US Bank or branch.
- ✓ Add $5.00 for Air Mail Subscription to Future *Adventures Unlimited* Catalogs

SPECIAL NOTES

- ✓ RETAILERS: Standard Discounts Available
- ✓ BACKORDERS: We Backorder all Out-of-Stock Items Unless Otherwise Requested
- ✓ PRO FORMA INVOICES: Available on Request
- ✓ VIDEOS: NTSC Mode Only. Replacement only.
- ✓ For PAL mode videos contact our other offices:

European Office:
Adventures Unlimited, Panewaal 22,
Enkhuizen, 1600 AA, The Netherlands
http: www.adventuresunlimited.nl
Check Us Out Online at:
www.adventuresunlimitedpress.com

Please check: ☑

☐ This is my first order ☐ I have ordered before ☐ This is a new addres

Name
Address
City
State/Province | Postal Code
Country
Phone day | Evening
Fax

Item Code	Item Description	Price	Qty	Total

Please check: ☑

- ☐ Postal-Surface
- ☐ Postal-Air Mail (Priority in USA)
- ☐ UPS (Mainland USA only)

Subtotal ➡
Less Discount-10% for 3 or more items ➡
Balance ➡
Illinois Residents 6.25% Sales Tax ➡
Previous Credit ➡
Shipping ➡
Total (check/MO in USD$ only) ➡

☐ Visa/MasterCard/Discover/Amex

Card Number
Expiration Date

10% Discount When You Order 3 or More Items!

Comments & Suggestions

Share Our Catalog with a Frienc